Land Allocation for Biomass Crops

Ruopu Li • Andrea Monti

Editors

Land Allocation for Biomass Crops

Challenges and Opportunities with Changing Land Use

 Springer

Editors
Ruopu Li
Department of Geography and
Environmental Resources
Southern Illinois University-Carbondale
Carbondale, IL, USA

Andrea Monti
Alma Mater Studiorum – University
of Bologna
Bologna, Italy

ISBN 978-3-030-09014-2   ISBN 978-3-319-74536-7   (eBook)
https://doi.org/10.1007/978-3-319-74536-7

Printed on acid-free paper

This Springer imprint is published by the registered company Springer International Publishing AG part of Springer Nature.
The registered company address is: Gewerbestrasse 11, 6330 Cham, Switzerland

Foreword

During the last 20 years, my research has evolved from field work with energy and industrial crops to modelling and assessment of the economic, environmental and social sustainability of biomass value chains. Resource efficiency and use of marginal lands for such crops form part of both my recent research area in respective collaborative projects and my teaching topics for graduate students of various disciplines. This book brings together knowledge from recent work with various lignocellulosic and herbaceous non-food crops, and addresses the challenges and opportunities of their cultivation in low quality and marginal lands. I strongly believe it brings high-added values to everyone interested in this field.

The editors, Ruopu Li and Andrea Monti, brought together a group of distinguished authors who have not only provided comprehensive documentation of research conducted for land use and biomass production, but also unique and valuable analyses of information that have not been published elsewhere.

The book content reflects recent developments, trends and advances in the area and includes a set of well-integrated, original research contributions that focus on state-of-the-art research. It addresses opportunities and challenges associated with the allocation of land resources to biomass cropping and offers valuable contributions to informed land management decision-making. The individual chapters are organized in response to cutting-edge questions associated with biofuels-related land-use issues in many countries such as Spain, Brazil, USA and China, and further provide informed and tailored insights.

Chapters 2, 3 and 4 address the availability of land, especially marginal categories, to accommodate future cultivation of lignocellulosic and other specialized herbaceous biofuel crops. More specifically, in Chap. 2, Sánchez et al. address two key challenges for bioenergy implementation: (i) 'Do we have land available to accommodate future biomass production?' (ii) 'What would be the choice of energy crops from the knowledge gained in field experiments?', with a specific focus in Spain. In Chap. 3, Parenti et al. evaluate possibilities and limitations for the cultivation of two promising perennial biofuel crops (giant reed and switchgrass) in Europe, in areas with natural constraints (ANC land). In Chap. 4, Chen et al. analyse the availability and economic feasibility of using marginal lands to cultivate five energy plant species, namely *Manihot esculenta, Jatropha curcas, Helianthus tuberous* L, *Pistacia chinensis* and *Xanthoceras sorbifolia* Bunge. Barbosa et al. analyse the opportunities of using contaminated land, for biomass production, in Chap. 5.

Chapters 6, 7 and 8 analyse the combination of employing diverse approaches (e.g. economic models and surveying) in three case studies with a focus on dynamics on farmers' planting decisions. More specifically, in Chap. 6, Granco et al. present the findings of a study on how Brazilian farmers decide which agricultural production to pursue and which land use to replace in the new frontier of sugarcane production. Varble and Secchi, in Chap. 7, examine the factors that are significant indicators in the interest of farmers to produce switchgrass through the analysis of the results of a survey completed by farmers in the Clear Creek watershed in rural Iowa. In Chap. 8, Kurkalova et al. evaluate land-use impacts of corn stover markets for the state of Iowa, as well.

Zhang et al., in Chap. 9, tackle the carbon sequestration capacity of farmland stover using a carbon capture model and land use and crop yield datasets in a region of North China. Finally, Dumortier, in Chap. 10, uses an economic model to identify the effects of plausible scenarios of cellulosic biofuel mandate on land allocation, nitrogen use and farmers' participation in US Conservation Reserve Program (CRP).

The book addresses a wide audience:

- Professors, graduate students and researchers that work with land-use modelling and field crop research: the relevant chapters offer comprehensive and transparent information on methodological approaches, field work and modelling, as well as a scientific sound discussion on the research results (robustness of the results, strengths and weaknesses of the approach, future research, etc.); and
- Policy and decision makers: the book includes thoroughly analysed evidence for the environmental, economic and social implications from using marginal land for biofuel crops as well as the dynamics of farmers' planting decisions.

In summary, this book is a milestone on the path to understand land use for biomass and biofuels production. It is a recommended reference for anyone who has an interest in such issues, regardless of whether, or not they have any experience or knowledge of land use modelling or field crop research for biofuels. Without question, I will use it in my research and highly recommend it to my students.

Imperial College London   Calliope Panoutsou
London, UK

Contents

Contributors

Pedro Luis Aguado Agro-Energy Group, College of Agricultural Engineering, Technical University of Madrid, Madrid, Spain

Parenti Andrea Department of Agricultural and Food Sciences, University of Bologna, Bologna, Italy

Bruno Barbosa Universidade Nova de Lisboa, Faculdade de Ciências e Tecnologia, Departamento de Ciências e Tecnologia da Biomassa, MEtRiCS, Caparica, Portugal

Universidade de São Paulo, São Paulo, Brazil

Jason Bergtold Kansas State University, Department of Agricultural Economics, Manhattan, KS, USA

Marcellus Caldas Kansas State University, Department of Geography, Manhattan, KS, USA

Lambertini Carla Department of Agricultural and Food Sciences, University of Bologna, Bologna, Italy

Yuqi Chen China Land Surveying and Planning Institute, Key Laboratory of Land Use, Ministry of Land and Resource, Beijing, China

Jorge Costa Universidade Nova de Lisboa, Faculdade de Ciências e Tecnologia, Departamento de Ciências e Tecnologia da Biomassa, MEtRiCS, Caparica, Portugal

María Dolores Curt Agro-Energy Group, College of Agricultural Engineering, Technical University of Madrid, Madrid, Spain

Jerome Dumortier School of Public and Environmental Affairs, Indiana University – Purdue University Indianapolis, Indianapolis, IN, USA

Allen Featherstone Kansas State University, Department of Agricultural Economics, Manhattan, KS, USA

Jesús Fernández Agro-Energy Group, College of Agricultural Engineering, Technical University of Madrid, Madrid, Spain

Ana Luisa Fernando Universidade Nova de Lisboa, Faculdade de Ciências e Tecnologia, Departamento de Ciências e Tecnologia da Biomassa, MEtRiCS, Caparica, Portugal

Gabriel Granco Stroud Water Research Center, Avondale, PA, USA

Xudong Guo China Land Surveying and Planning Institute, Key Laboratory of Land Use, Ministry of Land and Resource, Beijing, China

Qiaoli Hu Zhongke Haihui Technology Co., Ltd, Beijing, China

Lyubov A. Kurkalova Departments of Economics and Energy and Environmental Systems, North Carolina A&T State University, Greensboro, NC, USA

Ruopu Li Department of Geography and Environmental Resources, Southern Illinois University-Carbondale, Carbondale, IL, USA

Xiubin Li China Land Surveying and Planning Institute, Key Laboratory of Land Use, Ministry of Land and Resource, Beijing, China

Xingran Liu Key Laboratory of Agricultural Water Resources & Hebei Key Laboratory of Agricultural Water-Saving, Center for Agricultural Resources Research, Institute of Genetics and Developmental Biology, Chinese Academy of Sciences, Shijiazhuang, China

Chunyan Lv China Land Surveying and Planning Institute, Key Laboratory of Land Use, Ministry of Land and Resource, Beijing, China

Andrea Monti Alma Mater Studiorum – University of Bologna, Bologna, Italy

Javier Sánchez Agro-Energy Group, College of Agricultural Engineering, Technical University of Madrid, Madrid, Spain

Ana Cláudia Sant'Anna The Ohio State University, Department of Agricultural, Environmental, and Development Economics, Columbus, OH, USA

Silvia Secchi University of Iowa, Iowa, IA, USA

Yanjun Shen Key Laboratory of Agricultural Water Resources & Hebei Key Laboratory of Agricultural Water-Saving, Center for Agricultural Resources Research, Institute of Genetics and Developmental Biology, Chinese Academy of Sciences, Shijiazhuang, China

Dat Q. Tran Department of Agricultural Economics and Agribusiness, University of Arkansas, Fayetteville, AR, USA

Sarah Varble Southern Illinois University-Carbondale, Carbondale, IL, USA

Dengpan Xiao Institute of Geographical Sciences, Hebei Academy of Sciences, Shijiazhuang, China

Yucui Zhang Key Laboratory of Agricultural Water Resources & Hebei Key Laboratory of Agricultural Water-Saving, Center for Agricultural Resources Research, Institute of Genetics and Developmental Biology, Chinese Academy of Sciences, Shijiazhuang, China

Introduction

Ruopu Li and Andrea Monti

Abstract During recent decades, we have witnessed unprecedented expansion and intensification in land areas for biomass production around the world (Murphy et al. 2011; Taheripour and Tyner 2013; Taheripour et al. 2017). This worldwide pursuit for biofuels can be largely attributed to a thirst for cleaner fuels, global warming concerns and legislative mandates on transportation biofuels. Many countries have established policies to support and regulate the production and use of biofuels from biomass feedstocks (Havlík et al. 2011). For example, the U.S. Renewable Fuel Standards mandates at least 136 billion liter of liquid fuels to be blended into transportation fuels by 2022, and the current EU Renewable Energy Directive (RED) endorsed a mandatory target of a 10% share of biofuels in the transport petrol and diesel consumption by 2020. Moreover, to mitigate indirect land use change (iLUC) risks from dedicated non-food crops grown on existing agricultural land used for food and feed productions, the share of energy from biofuels produced from cereal and other starch-rich crops, sugars and oil crops and from crops grown as main crops primarily for energy purposes on agricultural land shall be no more than 7% of the final consumption of energy in transport in the EU in 2020. The U.S. Department of Energy (DOE)'s billion ton study (BTS) provides the most recent estimates of potential biomass that could be used for industrial biofuel production (U.S. DOE 2016).

R. Li (✉)
Department of Geography and Environmental Resources, Southern Illinois
University-Carbondale, Carbondale, IL, USA
e-mail: Ruopu.Li@siu.edu

A. Monti
Alma Mater Studiorum – University of Bologna, Bologna, Italy
e-mail: a.monti@unibo.it

1 Background

During recent decades, we have witnessed unprecedented expansion and intensification in land areas for biomass production around the world (Murphy et al. 2011; Taheripour and Tyner 2013; Taheripour et al. 2017). This worldwide pursuit for biofuels can be largely attributed to a thirst for cleaner fuels, global warming concerns and legislative mandates on transportation biofuels. Many countries have established policies to support and regulate the production and use of biofuels from biomass feedstocks (Havlík et al. 2011). For example, the U.S. Renewable Fuel Standards mandates at least 136 billion liter of liquid fuels to be blended into transportation fuels by 2022, and the current EU Renewable Energy Directive (RED) endorsed a mandatory target of a 10% share of biofuels in the transport petrol and diesel consumption by 2020. Moreover, to mitigate indirect land use change (iLUC) risks from dedicated non-food crops grown on existing agricultural land used for food and feed productions, the share of energy from biofuels produced from cereal and other starch-rich crops, sugars and oil crops and from crops grown as main crops primarily for energy purposes on agricultural land shall be no more than 7% of the final consumption of energy in transport in the EU in 2020. The U.S. Department of Energy (DOE)'s billion ton study (BTS) provides the most recent estimates of potential biomass that could be used for industrial biofuel production (U.S. DOE 2016).

Growing demand of this renewable fuels, driven by policy mandates, economic incentives and technical advancement in bioconversion techniques, has posed pressure and threats to existing agro-ecosystems. Almost all major resources (e.g., agricultural and forestry) of biofuels production are associated with accompanying environmental footprints (Hammond and Seth 2013; Hammond and Li 2016). The expansion of biomass cultivation is often at the cost of reduced land availability for dedicated food production and losses of land with critical ecological functions such as forests and wetlands. For example, spatial cropland datasets have shown cropland expands rapidly on grassland and other ecologically sensitive lands as a result of U.S. biofuels boom (Lark et al. 2015). Large-scale draining of prairie wetlands for ethanol-corn production has been observed in the US (Brooke et al. 2010). Besides, the net carbon benefits from biofuels are even under scientific scrutiny and policy debates due to the carbon emission associated with the direct and iLUC conversions (Searchinger et al. 2017; Fazio and Monti 2011). This land conversion process often involves complex interplay of physical dynamics and human systems that are driven by numerous geographic and socio-economic factors at different scales. Thus, state-of-the-art research on the land use issues surrounding the biomass production and its environmental impacts is important for informed land management decision making.

This book is edited and organized in response to the needs for more research in biofuels-related land use issues. This is probably among the first few books that are dedicated to this critical research topic. With contributions from international experts, this book seeks to establish a forum to explore a few cutting-edge questions associated with the land use and biomass production, such as 'do we have enough land, either primary or marginal, to accommodate future production of biomass?', 'how was farming decisions made in response to biomass incentives?', 'is the current biomass production socially, economically and environmentally sustainable?', and 'what are the main constrains currently limiting biofuels deployment?'

2 Overview of the Chapters

A total of ten chapters are included in this book. These chapters can be categorized into the following three general topics: (1) top-down land availability and its implications to bioenergy production, (2) bottom-up dynamics on farmers' planting decisions on cultivating energy crops, and (3) environmental effects and consideration associated with biomass development.

The first category includes four chapters (Chaps. 2, 3, 4, and 5). With a focus on land resources in Spain, Chap. 2 addresses two key challenges for bioenergy development: 'Do we have land available to accommodate future biomass production?', and 'What would be the choice of energy crops from the knowledge gained in field experiments?'. Chapter 3 evaluates the potentials and limitations of cultivating two promising perennial biofuel crops (giant reed and switchgrass) in Europe and their related natural resource constraints. In Chap. 4, the potentials of cultivating five specialized feedstock crops, including *Manihot esculenta*, *Jatropha curcas*, *Helianthus tuberous* L, *Pistacia chinensis*, and *Xanthoceras sorbifolia* Bunge, using marginal land in China were analyzed. Chapter 5 provides an in-depth review of less attended research issues in bioenergy literature, the opportunities and risks associated with production of energy crops on a specialized marginal land, i.e., contaminated land.

The second category has a collection of three chapters with a focus on dynamics on farmers' planting decisions (Chaps. 6, 7, and 8). Chapter 6 examines farmers' planting decisions on sugarcane production in two states of Brazil using a partial adjustment framework. Chapter 7 discusses the farmers' willingness to adopt switchgrass based on results of a survey conducted in an Iowa watershed, and it provides critical insights into bioenergy and land use policies. In Chap. 8, an economic model was employed to simulate farmers' profit-maximizing choices on crop-tillage rotations and stover collection, and evaluate the impacts of the stover collection restrictions on land use decisions and water quality implications.

The third category focuses on environmental effects associated with the biomass development, and spans across two chapters (Chaps. 9 and 10). Chapter 9 tackles the carbon sequestration capacity of farmland stover using a carbon capture model and land use and crop yield datasets in a region of North China. Chapter 10 uses an economic model to identify the effects of plausible scenarios of cellulosic biofuel mandate on land allocation, nitrogen use, and farmers' participation in U.S. Conservation Reserve Program (CRP).

3 Opportunities and Challenges

The chapters included in this book echoes various opportunities and challenges associated with land allocation for biomass development in current literature (e.g., Richard 2010; Hudiburg et al. 2016; Emery et al. 2017).

In view of negative effects from the development of conventional food-based biofuels, the advanced biofuels, developed based on non-food sources, provide unique opportunities of reducing carbon footprints in energy production (Valentine et al. 2012). To sustain a prospective 2nd-generation (2G) bioenergy industry, reliable and continuing supplies of biomass feedstock are critical. Chapters 2, 3, and 4 directly address the availability of land, especially marginal land, to accommodate future cultivation of lignocellulosic and other specialized herbaceous biofuel crops. Marginal and contaminated land offers a good opportunity for growing biofuel crops without competing with food land, but the matter is undoubtedly more complex than it first seems (Monti and Alexopoulou 2017). Chapter 5 discusses the opportunities of using contaminated land, a special type of marginal land, for biomass production. In addition, farmers' willingness and decisions to plant and harvest lignocellulosic biofuel feedstocks, typically sensitive to costs and profits, government incentives and policies, are crucial to establish a sustainable advanced fuel industry and potentially meet a government's biofuel mandate. The bottom-up dynamics and decision-making process occurred at the individual or farm scale need to be better understood for a viable 2G biofuel industry as well. Chapters 6, 7, and 8 tackles this interaction using different approaches (e.g., economic models and surveys) in three case studies. Lastly, positive environmental benefits are often regarded as important byproducts of 2G bioenergy development. For example, lignocellulosic crops usually require increased carbon sequestration and less fertilizer inputs. Chapter 9 presents a spatial assessment of carbon captured in corn stover, and Chap. 10 tackles the potential effects of cellulosic biofuel mandate on nitrogen use.

Meanwhile, there have been numerous challenges associated with the allocation of land resources for biomass planting. Challenges include limited land availability, indeterminate environmental effects (e.g., secondary land conversion, loss of habitats), and a growingly uncertain policy environment and global economy (Renzaho et al. 2017). For instance, primary arable land is prohibited from growing food-based energy crops in some countries such as China, whose policy only allows the cultivation on marginal land as demonstrated in Chap. 4. The limitation and policy barrier could affect the establishment of viable emerging market. In addition, potential environmental effects of planting biomass, especially iLUC conversion, needs to be carefully addressed (Khanna et al. 2017). Chapters 9 and 10 address these effects using environmental accounting based on current land use conditions and future scenarios. Furthermore, as current land modeling practice is predominantly scale-specific, either top-down or bottom-up, a cross-scale approach combining top-down and bottom-up mechanisms is critical and needed to modeling land availability and developing its scenarios in support of agricultural and biofuel policies (Li et al. 2013).

References

Brooke R, Fogel G, Glaser A, Griffin E and Johnson K (2010) Corn ethanol and wildlife – How increases in corn plantings are affecting habitat and wildlife in the prairie pothole region. [Online]. Available at: http://www.nwf.org/News-and-Magazines/Media-Center/Reports/Archive/2010/~/media/PDFs/Wildlife/01-13-10-Corn-Ethanol-Wildlife.ashx [February 27, 2013]

Emery I, Mueller S, Qin Z, Dunn JB (2017) Evaluating the potential of marginal land for cellulosic feedstock production and carbon sequestration in the United States. Environ Sci Technol 51:733–741. https://doi.org/10.1021/acs.est.6b04189

Fazio S, Monti A (2011) Life cycle assessment of different bioenergy production systems including perennial and annual crops. Biomass Bioenergy 35:4868–4878. https://doi.org/10.1016/j.biombioe.2011.10.014

Hammond GP, Li B (2016) Environmental and resource burdens associated with world biofuel production out to 2050: footprint components from carbon emissions and land use to waste arisings and water consumption. GCB Bioenergy 8:894–908. https://doi.org/10.1111/gcbb.12300

Hammond GP, Seth SM (2013) Carbon and environmental footprinting of global biofuel production. Appl Energy 112:547–559. https://doi.org/10.1016/j.apenergy.2013.01.009

Havlík P, Schneider UA, Schmid E et al (2011) Global land-use implications of first and second generation biofuel targets. Energy Policy 39:5690–5702. https://doi.org/10.1016/j.enpol.2010.03.030

Hudiburg TW, Wang W, Khanna M, et al (2016) Impacts of a 32-billion-gallon bioenergy landscape on land and fossil fuel use in the US. Nature Energy 1:nenergy20155. doi: https://doi.org/10.1038/nenergy.2015.5

Khanna M, Wang W, Hudiburg TW, DeLucia EH (2017) The social inefficiency of regulating indirect land use change due to biofuels. Nat Commun 8:15513. https://doi.org/10.1038/ncomms15513

Lark TJ, Salmon JM, Gibbs HK (2015) Cropland expansion outpaces agricultural and biofuel policies in the United States. Environ Res Lett 10:044003. https://doi.org/10.1088/1748-9326/10/4/044003

Li R, di Virgilio N, Guan Q et al (2013) Reviewing models of land availability and dynamics for biofuel crops in the United States and the European Union. Biofuels Bioprod Biorefin 7:666–684. https://doi.org/10.1002/bbb.1419

Monti A, Alexopoulou E (2017) Non-food crops in marginal land: an illusion or a reality? Biofuels Bioprod Biorefin 11(6):937–938. https://doi.org/10.1002/bbb.1820

Murphy R, Woods J, Black M, McManus M (2011) Global developments in the competition for land from biofuels. Food Policy 36:S52–S61. https://doi.org/10.1016/j.foodpol.2010.11.014

Renzaho AMN, Kamara JK, Toole M (2017) Biofuel production and its impact on food security in low and middle income countries: implications for the post-2015 sustainable development goals. Renew Sust Energ Rev 78:503–516. https://doi.org/10.1016/j.rser.2017.04.072

Richard TL (2010) Challenges in scaling up biofuels infrastructure. Science 329:793–796. https://doi.org/10.1126/science.1189139

Searchinger TD, Beringer T, Strong A (2017) Does the world have low-carbon bioenergy potential from the dedicated use of land? Energy Policy 110:434–446. https://doi.org/10.1016/j.enpol.2017.08.016

Taheripour F, Tyner W (2013) Biofuels and land use change: applying recent evidence to model estimates. Appl Sci 3:14–38. https://doi.org/10.3390/app3010014

Taheripour F, Zhao X, Tyner WE (2017) The impact of considering land intensification and updated data on biofuels land use change and emissions estimates. Biotechnol Biofuels 10:191. https://doi.org/10.1186/s13068-017-0877-y

Valentine J, Clifton-Brown J, Hastings A et al (2012) Food vs. fuel: the use of land for lignocellulosic "next generation" energy crops that minimize competition with primary food production. Glob Change Biol Bioenergy 4:1–19. https://doi.org/10.1111/j.1757-1707.2011.01111.x

U.S. Department of Energy (DOE) (2016) 2016 Billion-ton report: advancing domestic resources for a thriving bioeconomy, volume 1: economic availability of feedstocks. In: Langholtz MH, Stokes BJ, Eaton LM (Leads) ORNL/TM-2016/160. Oak Ridge National Laboratory, Oak Ridge, TN. 448p

Trends in Land Use in Spain and their Meaning to Bioenergy Production

Javier Sánchez, Pedro Luis Aguado, María Dolores Curt, and Jesús Fernández

Abstract The availability of land for biomass production has become a growing concern worldwide. Principles of sustainability indicate that the bioenergy chain, from biomass production to bioenergy end-uses, should be locally addressed. Thus, regardless of the possibilities that the international biomass trade could offer, each country should assess the size of the land resource that might be available for biomass production in order to establish realistic national targets for bioenergy. This work addresses two key challenges for bioenergy implementation: (i) 'Do we have land available to accommodate future biomass production?, (ii) 'What would be the choice of energy crops from the knowledge gained in field experiments?'. Focusing on Spain, an analysis of the evolution of land use is made and changes occurred in the last years are discussed to assess the size of land resource for bioenergy. Subsequently, the suitability of a selection of lignocellulosic energy crops is studied with a view to the potential available lands.

Keywords Land use · Lignocellulosic energy crops · Land suitability

1 Introduction

According to the World Bioenergy Association (WBA 2014), in the period 2000–2011 the bioenergy experienced the highest absolute increase among the renewable energies and reached 14.3% share in the gross final energy consumption (GFEC) of the World. In Europe, a steady increase in renewables has been recorded in line with the binding target of 20% renewables for 2020 (European Parliament 2009). The European Biomass Association (AEBIOM 2015) estimates show that bioenergy accounted for 61.2% of all renewable energies consumed in 2013 in Europe and that its main sector of application was heat (74.6%), followed by

J. Sánchez · P. L. Aguado · M. D. Curt (✉) · J. Fernández
Agro-Energy Group, College of Agricultural Engineering, Technical University of Madrid, Madrid, Spain
e-mail: md.curt@upm.es

© Springer International Publishing AG, part of Springer Nature 2018   7
R. Li, A. Monti (eds.), *Land Allocation for Biomass Crops*,
https://doi.org/10.1007/978-3-319-74536-7_2

transport (12.5%) and electricity (12.8%). Forestry, agriculture and waste provided 87%, 10% and 3% of the total primary energy supply of biomass at a global scale (WBA 2014). In Europe, about 30% of total bioenergy feedstock is based on agriculture and waste feedstocks (AEBIOM 2015). It is expected that the demand for bioenergy grows in Europe, which in turn will have to be met by an increase in biomass supply. Beyond the 'food vs. bioenergy' debate, the fact is that land is a limited resource and the increasing demand for bioenergy is generating land-use conflicts (Dauber et al. 2012).

As regards the land use in Europe, the most common primary land use is the agricultural one, with a share of 43% of total area; however, it varies widely from one country to another (minimum 7.4% in Finland to 73.2% in Ireland). Thus, Spain is above the average, with 52% total area (EUROSTAT 2015). Concerning bioenergy, agriculture in Europe provides feedstocks mainly for 1st generation transportation biofuels and also for biogas; additionally agricultural by-products and residues, such as straw, are used for heat and power. Taking Spain as a case-study, feedstocks of consumed bioethanol are corn (49.5% in 2011), sugarcane (24.8%) and wheat (18.0%); nearly 27% feedstocks came from crops grown in Spain. Most biodiesel consumed in Spain is produced from foreign feedstocks, Argentinian soybean (48%) and Indonesian palm oil (35%) (CNE 2013). Additionally, crop residues are also used for bioenergy; cereal straw is the feedstock of a number of power plants in Spain, e.g., the ACCIONA biomass power plants (61 MWe in all; http://www.acciona-energia.com/es). Based on these values, it is clear the Spain's dependence on foreign feedstocks of biofuels and the very little penetration of bioelectricity produced from feedstocks generated by agriculture.

The production chain of bioenergy, from biomass production to bioenergy end-uses, should be locally addressed in order to be sustainable. Recently the issue of securing that bioenergy comes from indigenous resources has been undertaken by Welfle et al. (2014a) for the UK. In another study (Welfle et al. 2014b), these authors categorized the biomass resources into: Grown Resource (food and non-food energy crops), Residues Resource (from forestry, industries & processes) and Waste Resource (waste wood, tertiary organic waste). In order to increase the biomass supply, the use of residues and waste resources for bioenergy should be optimized. As regards the grown resource, two major and inter-linked issues should be tackled: the availability of land for energy crops and the suitability of a particular energy crop to the available land. At a country scale, the former is related to the country's land use and country's policies whereas the suitability of a plant species is dependent on the country's agro-ecological conditions. For instance, some high-yielding biomass crops, like miscanthus, have shown great potential for some areas of North Europe but they may not be suitable for southern regions unless irrigation is provided (see Hastings et al. 2009).

Currently, European bioenergy policies move to other feedstocks than edible biomass resources, which have caused high controversy in the last decade on the grounds of food security and the effects of direct and indirect land use change (LUC, ILUC). In 2015 the European Parliament and the Council agreed 7% ceiling for conventional biofuels (i.e., those produced from cereals and other starch-rich crops, sugars and oil crops) and established that the energy contribution of lignocellulosic material (e.g. woody energy crops) and of non-food cellulosic material (e.g. energy crops such as ryegrass, switchgrass, miscanthus, giant cane) to the renewable energy target of Europe (10% of the final consumption of energy in transport by 2020) would be considered to be twice their energy content in energy reports (European Parliament 2015). If the focus of interest is the Grown Resource category (in the meaning by Welfle et al. 2014b), the fact is that non-food lignocellulosic energy crops are very limited in Europe but they are expected to expand (AEBIOM 2015). In this regard, great efforts have been made from the side of Research & Development (R&D) programmes, like the seventh Framework Programme (FP7) of the European Commission. For instance, perennial grasses for lignocellulosic biomass have been studied in the frame of the European projects known by the acronyms: GrassMargins, OPTIMISC and OPTIMA, and woody crops investigated in the project of acronym LOGISTEC. The reader can find details of these projects and many others on the subject, at the CORDIS European Commission portal, which is available at: http://cordis.europa.eu/projects/home_en.html.

In Spain, the suitability of dedicated energy crops for the Spanish conditions has been investigated at the national and regional (Robledo and Correal 2013; Sixto et al. 2013; AGAPA 2012) levels for several years. In all likelihood, the most relevant project at the national scale was the Singular and Strategic Project for the development, demonstration and evaluation of energy crop biomass-based energy production in Spain (acronym: PSE-On Cultivos, 2008–2012). In the framework of such project, a national network for biomass was set up between agricultural organizations, companies, research centers and end-users under the auspices of the Ministry of Economy and Competitiveness of Spain. Relevant to the current study, demonstrative and experimental fields of energy plant species (annual herbs, perennial herbs, woody crops) were established throughout Spain. Data of fields localization can be found in literature (Ciria et al. 2015b; Cañellas et al. 2012; Pérez et al. 2014; Djomo et al. 2015; Carrasco et al. 2014; Sanz et al. 2011a, 2011b; Maletta et al. 2016). The experience gained on energy crops over the years in Spain set the basis for further studies.

The issue of which land could be allocated to sustainably produce dedicated energy crops has been of major concern since land conversion for biofuel feedstock production may have environmental and social consequences. On the one hand and, as regards to agricultural areas, cultivation of bioenergy crops on land devoted to food production should be avoided in order to ensure food security and stable food markets. Whereas on the other hand, well-adapted energy crops grown in surplus agricultural land, abandoned farmland or marginal land could contribute to meet the European and national targets on renewable energy consumption, while protecting soil from erosion and further degradation and providing farmers with economically-viable alternatives against rural areas depopulation. Therefore, understanding land use and land use statistics is essential to assess the agricultural land potentially available for the suitable and sustainable production of biomass resources for bioenergy purposes.

Specifically, our study addresses two key challenges for bioenergy implementation in Spain: the availability of land to accommodate future biomass production and their suitability for growing energy plant species. Accordingly, an analysis of the evolution of land use is made and changes occurred in the last years are discussed to assess the size of land resource for bioenergy. Subsequently, the suitability of a selection of lignocellulosic energy crops is studied with a view to the potential available lands. The novelty of this work lies in the application of an integrated approach for Spain by which land-use statistical data, field work data and georeferenced information are jointly analyzed in order to achieve the objectives pursued.

2 Material and Methods

2.1 Patterns of Land Use and Land Use Change in Spain

Statistics on the land use in Spain at NUTs3 (Nomenclature of Territorial Units for Statistics of the European Union at the region level) scale were collected for the time period 1998–2013 from the Annual Agricultural Statistics (AAS) published by the Ministry of Agriculture, Food and Environment of Spain (MAGRAMA 2015). Among other data, these yearbooks gather official statistics on the main agricultural land uses following the classification shown in Table 1. This information was compiled and harmonized by year and NUTs3 scale (province) for building a geo-referenced database for the 15-year period in order to analyze its evolution both, temporally and spatially.

The agricultural land considered in our work as potentially available for bioenergy - i.e. for growing dedicated energy crops adapted to the specific conditions of Spanish regions- was the fallow land as well as the abandoned arable land (see Allen et al. 2014), in both, rain-fed and irrigated conditions.

Table 1 Land use classes according to AAS (MAGRAMA 2015)

Agricultural land	Arable land	Herbaceous crops	Rain-fed
			Irrigated
		Fallow land	Rain-fed
			Irrigated
	Permanent crops		Rain-fed
			Irrigated
Land mainly use for grazing		Natural grasslands	
		Pastures	
		Uncultivated/barren land	
Forest land			
Other land uses			

Fallow land is defined in the Council Regulation 543/2009 as all arable land included in the crop rotation system, whether worked or not, but with no intention to produce a harvest for the duration of a crop year. It could be considered as abandoned farmland, the piece of land uncultivated or left fallow for a minimum of five consecutive years if it is not used for grazing (permanent meadows and pastures) or overgrown with trees (forest or other wooded land (FAO 2014)).

Motives leading farmers to leave their arable land fallow could be basically grouped into five overlapping causes: agronomic-environmental, economic, human, institutional and political (Sauer et al. 2012). Within these causes, some specific factors could be detailed, e.g. crop rotation, soil type and quality, plot isolation, low economic profitability, low market share or under-developed market of agricultural products, lack of equipment and manpower and/or policy framework. In any event, fallow land is recorded in the official statistics of Member States (MS), especially after the greening action of the Common Agricultural Policy (European Parliament 2013), by which the fallow land could count as a diversification practice (Ecological Focus Areas – EFAs) for the greening payment. As already mentioned, data on fallow land at NUTs3 scale is compiled in the AAS yearbooks.

Similar to land lying fallow, other land use changes of agricultural areas over the years like farmland abandonment, respond to several inter-related factors regarding economic, political, social and environmental drivers. In the European Union (henceforth EU), the main drivers leading to farmland abandonment could be attributed to weak land markets, low farm income, low density population and remoteness, as well as lack of investments and/or farm-holder's age (Terres et al. 2013). In contrast to fallow land, abandoned agricultural land is not tracked or accounted in official datasets.

In order to assess the abandoned arable land in Spain, the geo-referenced databases from the CORINE Land Cover inventory (henceforth, CLC) for the time series 2000 (CLC2000), 2006 (CLC2006) and 2012 (CLC2012) were compiled (http://land.copernicus.eu/pan-european/corine-land-cover/view). In this geo-database, land use and land cover changes (change layers) are identified by visual interpretation of high resolution satellite imagery. In the current work, it was assumed that the abandonment of arable land is reflected in the change layers among the land use classes shown in Table 2 -i.e. those polygons which were classified as arable (2.1.1 or 2.1.2) in CLC2000 and then identified as naturally-driven vegetated areas (classes 2.4.3, 3.2.1, 3.2.3, 3.3.3 and 3.3.4) in CLC2006 and CLC2012 due to the lack of agricultural operations.

From these layers, the percentage of the arable land reduction due to farmland abandonment (as explained above) over the total arable land diminution (which would also include changes due to cultivation of permanent crops, urbanization, industrialization, etc.) was calculated at NUTs3 scale. Then, in those regions where a reduction in the arable land area in the period 2000–2013 was recorded in the AAS, the previously calculated percentage was applied to the compiled statistics in order to estimate the decrease in arable land driven by the abandonment of cultivated areas.

Table 2 Land use classes for assuming arable land abandonment according to the Corine Land Cover (CLC) classes in the time series 2000 (CLC2000), 2006 (CLC2006) and 2012 (CLC2012)

CLC2000 class	CLC2006 class	CLC2012 class
2.1.1. Non-irrigated arable land	2.4.3. Land principally occupied by agriculture, with significant areas of natural vegetation	2.4.3. Land principally occupied by agriculture, with significant areas of natural vegetation
2.1.2. Permanently irrigated land	3.2.1. Natural grassland	3.2.1. Natural grassland
	3.2.2. Moors and heathland	3.2.2. Moors and heathland
	3.2.3. Sclerophyllous vegetation	3.2.3. Sclerophyllous vegetation
	3.3.3. Sparsely vegetated areas	3.3.3. Sparsely vegetated areas
	3.3.4. Burnt areas	3.3.4. Burnt areas

With the aim of assessing the land potentially available for bioenergy with no distortion of the food production chain, two scenarios were considered: Sc-I (50% of fallow land + abandoned agricultural land) and Sc-II (100% of fallow land + abandoned agricultural land). Thus, land left fallow as a crop rotation practice was taken into account in the former scenario.

2.2 Land Suitability Assessment for Growing Energy Crops

Land suitability for a number of lignocellulosic energy crops was assessed following a multi-criteria decision analysis by which crop agro-ecological requirements are compared to land agro-climate conditions in a geographical information system (GIS) environment. The approach was specifically developed for a Spain-funded project (PSE-On Cultivos) and was based on the agro-ecological zoning guidelines developed by FAO (FAO 1996; IIASA/FAO 2012).

The methodology comprised four steps: (i) Selection of target energy crops, (ii) Definition of soil & climatic requirements of the target energy crops, (iii) Compilation of georeferenced information; (iv) Spatial analysis for agro-ecological zoning. Tools provided by ArcGIS 10.0 (ESRI™) mapping software were used for all GIS works.

Selection of Target Energy Crops

The range of plant species used/proposed as dedicated energy crops is very wide (El Bassam 2010), including amylaceous, oil and lignocellulosic plant species and algae (macro- and micro-algae). Concerns on food security and land use change (LUC, ILUC) have led to the promotion of non-food cellulosic feedstocks for bioenergy, like grassy energy crops and woody energy crops (Directive 2015/1513 of the European Union) grown as short rotation coppice (SRC). Hence, the first criterion of selection for our work was the lignocellulosic nature of plant species. Other criteria were the existence of species background information in Spain, since the potential of a particular plant species for bioenergy is site-dependent, and proved hardiness. So, lignocellulosic and hardy plant species already experimented in Spain or in similar climates were considered for our study.

Crop Soil and Climatic Adaptability Inventory

Soil and climatic adaptability inventory for the target plant species was based on the Ecocrop database (EcoCrop 2013), which was used both as a model and as a source of information. Other sources of information were the Plants Database (USDA and NRCS 2016), El Bassam (2010), Venendaal et al. (1997) and relevant literature on their cultivation. Whenever available, data from experiments in Spain or in the Mediterranean region were compiled. Besides, the records of the presence of the target species in Spain, as reported by the database of plant species of Spain (ANTHOS 2012–2016) were georeferenced and subsequently processed (see next Section) to refine the information on soil and climate requirements of the target plant species.

In our work, the dataset intended for each plant species involved ten variables; some variables were included in the analysis as binary -i.e. they may take only two values that represent success or failure-, whereas others were introduced as restrictive -i.e. they gradually affect plant growth according to established value ranges-. Concerning climatic variables the mean annual/monthly temperature (restrictive), minimum temperature (binary), mean annual/monthly precipitation (restrictive) and number of consecutive frost-free days (binary) were considered for the database of plant species requirements; concerning soil variables, the pH, depth and texture were incorporated to the model as restrictive variables while salinity and susceptibility to flooding were applied as binary variables. Additionally, land orientation (restrictive) was considered as well.

Georeferenced Information

In order to create GIS layers on the variables: mean annual temperature, monthly temperature, minimum temperature, annual precipitation, monthly precipitation, number of consecutive frost-free days, soil pH, depth, texture, salinity, susceptibility to flooding, drainage, and land use in Spain, georeferenced information was compiled. Databases used in this step and the respective resolution (pixel size) are given in Table 3. The fact of working with databases of different resolution does not represent an obstacle for undertaking a multi-criteria analysis; however, it entails limitations for the results of the analysis. For example, local changes in

Table 3 Sources of georeferenced information

Variable	Resolution	Source of information
Mean annual and monthly Temperature	1 km	AEMET (2012)
Minimum temperature		
Annual and monthly precipitation		
Number of consecutive frost-free Days		
Soil pH	5 km	Panagos et al. (2012), JRC (2010)
Soil depth		Panagos et al. (2012), Panagos (2006) y van Liedekerke et al. (2006).
Texture		
Salinity		Panagos et al. (2012), Toth (2008).
Susceptibility to flooding		Own approach based on data by AEMET (2012) and FAO/IIASA/ISRIC/ISS-CAS/JRC (2008)
Drainage	1 km	FAO/IIASA/ISRIC/ISS-CAS/JRC (2008)
Available water storage capacity		
Land use	100 m	EEA (2016)

soil properties may not be recorded in GIS layers and so, they may not be taken into account for the land suitability assessment. In this regard, the results from our assessment should be taken with caution.

Due to the fact that georeferenced information on land's susceptibility to flooding was not available, it was developed for this work according to the following approach:

(i) Calculation of monthly water balance: it was calculated according to a simplified method from the GIS layers of precipitation (P), potential evapo-transpiration (ETP) and available water capacity (AWC), using the application 'Model builder' in ArcGIS 10.0™ software.

Water surplus (Q_i) = Inputs of water$_i$ – Losses of Water$_i$
Inputs of water = $AWC_{i-1} + P_i$
Outputs of water = $ETP_i + AWC_i$
i = month

(ii) Creation of GIS surplus water layer: Cells exhibiting n-month continuous water surplus ($n = 1$–10 months) –based on GIS water balance layer calculated in the previous step- were extracted in order to create the surplus water layer.

(iii) Creation of GIS water sink layer: It was created over the digital extract model using Arc Hydro tools available in ArcGIS 10.0™ software, DEM extract elevations, from the data layers of surplus water (from the previous step) and drainage (FAO/IIASA/ISRIC/ISS-CAS/JRC 2008). Points enclosed within cells of higher elevation might act as water sink areas as long as the conditions of surplus water and poor soil drainage are met (Table 4). In that case, they would be categorized as areas susceptible to flooding.

Table 4 Conditions to be met by water sink areas. Drainage classes according to FAO/IIASA/ISRIC/ISS-CAS/JRC (2008)

Drainage class	Sink	Number of months with surplus water
Poor-very poor	Yes	> 1
Imperfectly	Yes	> 5
Other	No	≠

Spatial Analysis of Land Suitability for Growing Energy Crops

Land-suitability for the target energy crops was assessed following a GIS-based multicriteria decision analysis -based on attributes of soil and climate- using ArcGIS reclassification tools and map algebra. For binary variables the site condition was reclassified with the value '0' if it was not tolerated by the crop; on the contrary, the value '1' meant that it was suitable for the crop. Binary variables (see Sect. Crop Soil and Climatic Adaptability Inventory) were: minimum temperature tolerated by the crop, minimum number of frost free days in a year, salinity and flood tolerance. The other variables were reclassified with values categorized into 'not suitable' (value = 0), 'acceptable' (value = 1) and 'optimum' (value = 2), where 'acceptable' meant that the variable was within the range tolerated by the crop, and 'optimum' meant that the variable was within the range identified as optimal for the crop.

In order to perform the geoprocessing, *ad-hoc* models were built with the application ArcGIS Model Builder specifying variable values of each crop (see Fig. 1). Variables were combined by multiplying the layers that resulted from the variable reclassification above described. Cells with a value = 0 meant that the crop could not be grown at that site while cells with high values represented sites where the crop could thrive. No weighting among variables was applied in the model. The resulting layer was reclassified according to Quantile classification with five classes (quintiles), representing five classes of soil & climate suitability: not suitable, low suitability, moderate suitability, high suitability (=suitable) and very high suitability.

Finally, restrictions of land use were applied using Corine Land Cover 2012 inventory, in order to complete the agro-ecological zoning. In our approach, crops that can be grown in the rain-fed conditions of Spain were restricted to dry arable lands; conversely, high-water-demanding crops were restricted to irrigated arable lands.

3 Results and Discussion

3.1 Patterns of Land Use and Land Use Change in Spain

The statistics on agricultural land compiled from AAS from MAGRAMA (MAGRAMA 2015) showed that the total agricultural area at national scale

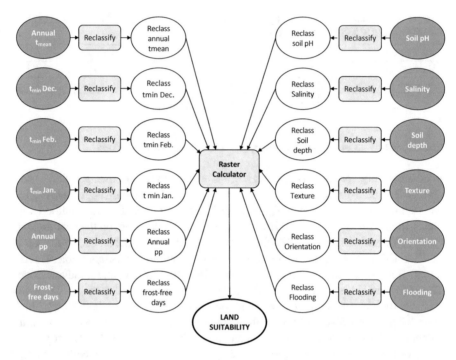

Fig. 1 Flow diagram of soil and climate suitability spatial analysis

decreased from 18.5 Mha in 1998 to 17.1 Mha in 2013. This declining of the overall agricultural area is also recorded in the rest of the EU and is expected to continue (Hart et al. 2013).

The evolution of the main land use classes of agricultural land in Spain for the period 1998–2013 is depicted in Fig. 2. In the overall statistics for Spain, the arable land with annual crops showed a decrease in both, rain-fed and irrigated conditions. In the former, the reduction amounted to 12.7% (from 7.97 to 7.07 Mha) whereas for the latter, this decrease was 13.3% (from 2.30 to 2.03 Mha).

At NUTs3 scale, a reduction in the rain-fed arable land with herbaceous crops occurred in 42 (84.0%) provinces, whereas in the case of irrigated arable land, the decrease was recorded in 37 (74.0%) provinces (Fig. 3). The reduction of the area devoted to annual crops in both, rain-fed and irrigated conditions is concentrated in the southern half of the Iberian Peninsula, where the province of Seville showed the largest decreases (nearly 200,000 ha and 26,000 for rain-fed and irrigated land, respectively).

On the contrary, the dry-land left fallow in the total figures for Spain remained steady over the analyzed years with an average area of 3.14 Mha, whereas the irrigated fallow land showed an increase of 231,107 ha from 1998 to 2013, with a mean value of 331,521 ha (Fig. 2). However, in both cases, a reduction in the fallow land was recorded in the last two years of the analyzed period (2012–2013), with a decrease of 15.0% and 11.1% for rain-fed and irrigated fallow

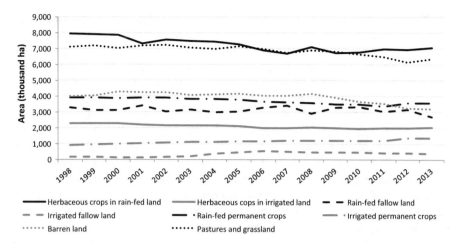

Fig. 2 Evolution of the agricultural land in Spain in the time period 1998–2013. (Source: AAS yearbooks from MAGRAMA 2015)

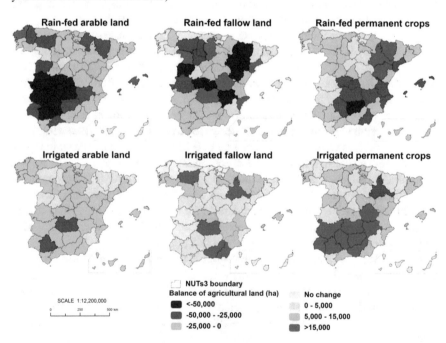

Fig. 3 Balance of agricultural land in Spain at NUTs3 scale in the time period 1998–2013. Negative values indicate a decrease in the area for each category

land, respectively. The average rainfall for Spain in the agricultural year of 2013 (September–August) was 806.8 mm, 24.1% higher than the mean value in the period 1971–2000 (AEMET 2014), which may be one of the main driving causes for the reduction of fallowing practices in Mediterranean areas since in these conditions soil water content increases and farmers expect higher yields.

The geographical representation of the land balance (Fig. 3) for these categories did not show a clear trend in land use change over the provinces of Spain, especially in the case of fallow dry-land. Nevertheless, it is worth noting that the growth of irrigated fallow land is widespread in most provinces (90%) of Spain. Thus, the total fallow land in 2013 in Spain amounted to 2,703,260 ha and 372,821 ha in rain-fed and irrigated conditions, respectively.

In the light of the compiled statistics, the abolition of the set-aside obligation (Regulation (EEC) 1272/88) by the CAP 'Health Check' reform in 2008, did not lead to a reduction of fallow land in Spain. By that regulation, since 1992 larger producers had to set aside a proportion of arable land and leave it temporarily uncultivated as a measure of reducing food surpluses at a time of high levels of cereal production. Within the analyzed time period in Spain, fallow land represented 27.1% of total arable land on average, ranging from 24.0% (2000) to 30.9% (2007). Hence, fallowing agricultural land in Spain seems not to be driven by regulations of food markets but by traditional practices for adapting agricultural production to biophysical constrains (low rainfall and soil fertility) and consequently to low profitability in such conditions. Therefore, an alternative for this untapped land may be of interest to farmers with no ensuing disruption of food production.

Regarding the agricultural area with permanent crops in Spain, the dry-land was reduced by 9.7% (346,293 ha) whereas the area under irrigation increased by 62.0% (231,107 ha). This fact is mainly driven by the conversion of rain-fed into irrigated agriculture of olive groves and vineyards. In the period 1998–2013 the olive groves under irrigation increased from 256,658 to 606,147 ha whereas the area of vineyards rose from 91,333 to 215,074 ha. On the other hand, the land devoted to olive production in rain-fed conditions decreased from 2.09 to 1.90 Mha while non-irrigated vineyards were reduced from 1.07 to 0.73 Mha. As it can be seen in Fig. 3, this land conversion mainly occurred in central and southern regions of the Iberian Peninsula, mostly in Castile-La Mancha in the case of vineyards (where rain-fed area decreased from 554,577 to 315,357 ha whereas irrigated vineyards increased from 39,261 to 126,649 ha) and in Andalusia for olive groves (where land devoted to olive production in rain-fed conditions decreased from 1,177,168 to 968,934 Mha while, under irrigation, it increased from 187,951 to 473,664 ha).

Independently to which land use the agricultural area has been allocated to, it is believed that the reforms of the Common Agricultural Policy (CAP) on decoupling payments from crop production could lead to several risks such as agricultural land abandonment (Renwick et al. 2013). Actually, the number of farms decreased in the EU-27 by about 9% in the period 2003–2007 (European Commission 2010; Raggi et al. 2013).

According to the assessment of agricultural land abandonment based on CLC databases, the share of arable land reduction due to farmland left aside, i.e. land use change from arable land into naturally-driven vegetated areas in the period 2000–2012, resulted in 1.4% and 5.9% for non-irrigated and irrigated land, respectively. This percentage was calculated for each NUTs3 in Spain, ranging from 0% to 17.3% in the case of rain-fed land and up to 26.9% for originally irrigated arable land.

The distribution of the abandoned land assessed in this study is shown in Fig. 4.

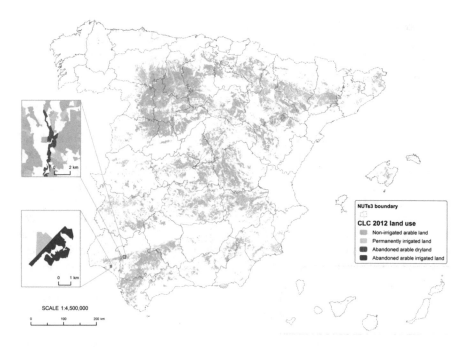

Fig. 4 Map of agricultural land classes according to CLC and the estimated arable land abandoned. Examples of detailed plots of estimated abandoned farmland are shown on the left side

As a result of applying the abovementioned specific percentage at NUTs3 scale over the statistics on arable land decrease (from AAS data), the estimated abandoned farmland amounted to 18,164 ha in the case of non-irrigated land and 15,885 ha for irrigated land.

The estimated abandoned arable dry land was particularly significant in Extremadura and Castile-La Mancha accounting to 5,101 and 4,550 ha, respectively. The high risk of farm abandonment in these NUTs2 regions in Spain was also pointed out by Terres et al. (2013), who estimated a composite risk index based on normalized values of individual drivers of farmland abandonment. The index calculated for the abovementioned regions was above 0.7 from a range between 0 and 1 (Terres et al. 2013). The main socio-economic drivers of farmland abandonment identified in Spain were remoteness and low density population, low investment ratios as well as low paid rent. Hotpots of abandoned farmland in Southwest of the Iberian Peninsula were also observed in the study of Estel et al. (2015) which was based on remote sensing tools on dense time series of satellite imagery.

Land taken out of agricultural production may be caused by the progression of land marginalization originated by several drivers (socio-economic, environmental or political) or it may be part of a rotational process by which the farmer leaves productive land set-aside (Lavalle et al. 2011). In the latter, some agricultural operations are carried out ensuring a minimum level of land maintenance in order to avoid habitats deterioration so as to be included as ecological set aside or ecological focus area (CAP greening). In the former, no human intervention is provided and so biodiversity may be positively or negatively affected: some semi-natural habitats and associated species could be damaged or threaten whereas opportunities for restoration of non-agricultural habitats (re-wilding) in fragmented landscapes may be generated (Keenleyside and Tucker 2010). Other studies pointed that some bioenergy crops, especially perennials (miscanthus and SRC, among others), could enhance biodiversity in comparison with arable production of annual crops, where mammals and birds species are respectively five and four times more numerous than in a control crop of wheat (Aylott and McDermott 2012) and an index of pollinator abundance could increase by an average of 11% (Meehan et al. 2013). The current study aimed at assessing different species of bioenergy crops in order to provide evidence of the possibilities of implementing diverse, heterogeneous and integrated plantations of dedicated species in the surplus agricultural land not devoted to food production for enhancing biodiversity and avoiding large extensions of monocultures.

Considering the fallow and abandoned farmland as the area potentially available for growing energy crops, the land for bioenergy in Spain amounts to 1,572,090 ha in scenario I (Sc-I = 50% fallow land + abandoned farmland) and 3,110,130 ha in scenario II (Sc-II = 100% fallow land + abandoned land). The contribution of each land use category is shown in Table 5. Non-irrigated fallow land represents 86% of the total potential area whereas irrigated fallow land accounts for 12%. In Sc-I, the contribution of the abandoned farmland to the potential bioenergy area amounts to 1.2% and 1.0% in rain-fed and irrigated conditions respectively, while in Sc-II it is reduced to 0.6% and 0.5%, respectively.

The distribution of the bioenergy potential area at NUTs3 scale is depicted in Fig. 5.

From the NUTs3 assessment, the contribution of each land use category to the potential bioenergy area is aggregated at NUTs2 scale (for a better visualization) in Figs. 6 and 7 for Sc-I and Sc-II, respectively. It can be noticed that in both scenarios, the land potentially available for bioenergy according to this study is mainly located in central Spain, being Castile-La Mancha (9) and Castile-Leon (14) the largest NUTs2 contributors with 545,261 and 319,262 in Sc-I and 1,079,802 ha

Table 5 Area potentially available for bioenergy per land use category for scenario I (Sc-I = 50% fallow land + abandoned farmland) and scenario II (Sc-II = 100% fallow land + abandoned land). Values in ha

Land use category	Sc-I	Sc-II
Rain-fed fallow land	1,351,630	2,703,260
Abandoned agricultural dry-land	18,164	18,164
Irrigated fallow land	186,411	372,821
Abandoned irrigated agricultural land	15,885	15,885
Total Spain	1,572,090	3,110,130

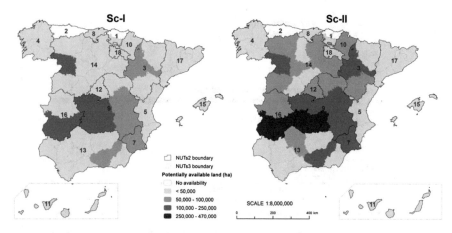

Fig. 5 Agricultural land potentially available for growing energy crops in scenario I (Sc-I = 50% fallow land + abandoned farmland) and scenario II (Sc-II = 100% fallow land + abandoned land). Numbers represent the NUTs2 codes as in Fig. 6 and Fig. 7

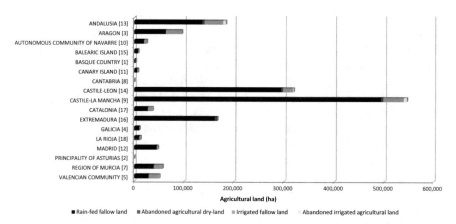

Fig. 6 Land potentially available for bioenergy in Sc-I (Sc-I = 50% fallow land + abandoned farmland) at NUTs2 scale, broken down by land use class. Between brackets, NUTs2 code as in Fig. 5

and 633,577 ha in Sc-II, respectively. The large dry-land area left fallow in these regions represent up to 31% and 18% of the total potential, respectively. As regards to abandoned farmland, its contribution to the bioenergy potential area is especially significant in Castile-La Mancha (9) and Andalusia (13) amounting to 10,719 and 10,025 ha, respectively.

In Fig. 8, the percentage of the agricultural land potentially allocated to bioenergy in relation to the total arable land for each NUTs2 in both scenarios is illustrated. Except for Principality of Asturias (2) where no potential area devoted to energy crops was estimated, ranges in Sc-I varied from 1.7% to 29.3% for Cantabria

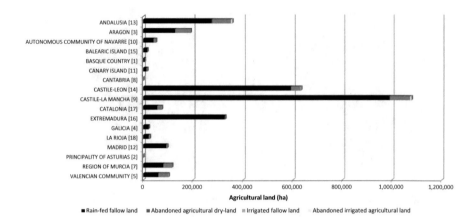

Fig. 7 Land potentially available for bioenergy in Sc-II (Sc-II = 100% fallow land + abandoned farmland) at NUTs2 scale, broken down by land use class. Between brackets, NUTs2 code as in Fig. 5

(8) and Valencian Community (5), respectively and from 3.4% to 58.1% in Sc-II for the same regions. Thus, regions where farmland lying fallow is a common agricultural practice for different reasons (low rainfall and limited soil content on organic matter, among others) are easily identified (e.g. Region of Murcia, Canary Island or Madrid). Hence, growing energy crops adapted to these specific agro-climatic conditions means a promising alternative for farmers in these regions.

Besides, devoting the fallow and abandoned farmland to bioenergy purposes would not distort the food production and food markets while offering an alternative for farmers and ensuring the supply of a renewable source of energy and bio-based products.

In terms of environmental protection of areas with high biodiversity value, a balanced mixture of the different species proposed in the current work with already established food-based crops would contribute to a more diverse agricultural environment whereas soil protection of sparsely vegetated areas (without a vegetative cover) from erosive agents would be achieved.

3.2   Land Suitability for Growing Energy Crops

Lignocellulosic plant species that have been grown or experimented as energy crops and that have shown good adaptation to some Spanish areas or to other regions with similar climates, were chosen for our study. Thus, lignocellulosic crops for bioenergy have been experimented in Spain for several years in the framework of European R + D + i projects -Cynara network, AIR3-CT93-1089; Biomass Sorghum, FAIR3-CT96 1913; BioEnergy Chains, ENK6-2001-00524; ECAS, Interreg III; Biocard; Optima, 7FP; and others- and Spanish projects like

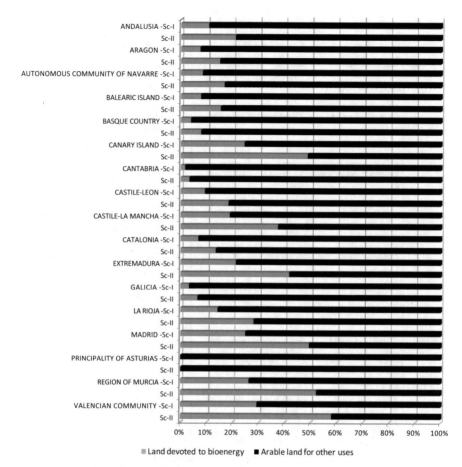

Fig. 8 Share of land devoted to bioenergy production in relation to total arable land at NUTs2 scale in scenarios I (Sc-I = 50% fallow land + abandoned farmland) and II (Sc-II = 100% fallow land + abandoned farmland)

the strategic project of acronym PSE-On cultivos and Probiocom, RTA2012-00082-C02-02. Plant species for this study were selected according to previous information (Table 6). Mention to the typology of crop management (rain-fed or irrigated) in Spain and to a selection of bibliographic references relevant to the case-study of Spain is also made.

Data on climate and soil requirements of the target plant species from different sources of information were compiled and summarized to facilitate the subsequent GIS-based multicriteria decision analysis. The resulting data sheet of key values for our study is shown in Table 7.

The challenge for biomass in Spain is the search of energy plant species able to withstand dry conditions. Most arable lands in Spain (81%, about $9.655*10^6$ ha according to Corine inventory) are classified as non-irrigated arable lands and

Table 6 Target energy crops

Habit	Botanical name	Common name	Crop management	Source of information
Perennial herbs	*Agropyron cristatum* (L.) Gaertn + relatives	Wheatgrass	Rain-fed	Utrilla et al. (2007); Piquero et al. (2011); Ciria et al. (2015a)
	Arundo donax L.	Giant reed	Irrigated	Christou et al. (2000); Curt (2009a); Dalianis and El Bassam (2010); Cosentino et al. (2014); Sánchez et al. (2016)
	Cynara cardunculus L.	Cardoon, cynara	Rain-fed	Fernández et al. 2006; Fernández (2009); Gominho et al. (2011); Curt et al. (2012).
	Miscanthus spp	Miscanthus	Irrigated	Bao et al. (1996); Walsh and Jones (2000); Lewandowski et al. (2003); Qingguo and El Bassam (2010)
	Panicum virgatum L.	Switchgrass	Irrigated	Elbersen (2002); Lewandowski et al. (2003); McLaughlin and Kszos (2005); Curt (2009b).
	Phalaris arundinacea L.	Reed canary grass	Irrigated	ADEME (1998); Andersson and Lindvall 2002; Lewandowski et al. (2003).
Annual herbs	*Secale cereale* L.	Rye (biomass)	Rain-fed	Sanz et al. (2011b); Ciria et al. (2015b)
	Sorghum bicolor (L.) Moench.	Sorghum (biomass)	Irrigated	Curt et al. (2001); Habayarimana et al. (2004); Curt et al. (2011).
	x Triticosecale Wittm. Ex A. Camus	Triticale (biomass)	Rain-fed	Sanz et al. (2011b); Ciria et al. (2015b)
Woody species	*Eucalyptus* spp.	Eucalyptus	Rain-fed	Pereira et al. (1990); Montoya (1995); Ceulemans et al. (1996).
	Populus spp	Poplar	Irrigated	Montoya (1988); Ciria (2009); Sixto et al. (2010 & 2013); Pérez (2016)
	Robinia pseudoacacia L.	Black locust	Rain-fed	Baldelli (1996); Montoya and Mesón (2004).
	Salix spp	Willow	Irrigated	Montoya and Mesón (2004); Facciotto et al. (2008);
	Ulmus pumila L.	Siberian elm	Rain-fed	Iriarte (2008); Fernández et al. (2009); Pérez (2016).

rainfall is usually low across Spain, 665 mm/year on average (series 1940–2010), although great differences between locations can be existent. Thus, it ranges from 119 mm/year in Fuerteventura (Canary Islands) to 1,534 mm/year maximum in Galicia Coast (NW coastal strip of the Iberian Peninsula). The driest month is July, generally coinciding with the main plant growth period whereas December is the rainiest one (MAGRAMA 2016a). Some well-known and high yielding energy crops, like miscanthus, switchgrass or giant reed, that can be grown rain-fed in

Table 7 Simplified data sheet of climatic and soil variables for target plant species in Spain

Target energy crop	Climatic variables			Soil variables											
	Mean temperature (tm, °C)		Minimum temperature (tmin, °C)	Rainfall (mm)		pH		Soil depth		Texture		Salinity (dS/m)		Flooding	Other
	Absolute	Optimal	Absolute	Absolute	Optimal	Absolute	Optimal	Absolute	Optimal	Absolute	Optimal	Absolute	Optimal		
Wheatgrass	5–45	15–25	Jan tmin > −4	200–800	250–500	4–8.5	5.5–6.5	Shallow, Deep	Medium	Fine, Medium – Fine	Medium, Coarse	Medium	Low	Moderately tolerant	
Giant reed	5.9–19.3	24 < tm < 30	Jan tmin > −6.5	230–1950	470–920	3.6–8.7	5.2–7.3	Shallow	Medium – Deep	Medium, Coarse	Fine, Medium to Fine	High	Low	Tolerant	
Cardoon	5.9–38	14–28	Jan. tmin> −6.5. Sowing: if autumn sowing: Sep–Oct tmin > 7; if spring sowing → Mar–May tmin> 7	350–1000	450–800	4.3–7.8	5.5–7.3	Medium	Deep	Fine, Medium – Fine	Coarse, Medium	Medium	Low	Non-tolerant	Preference for calcic soils
Miscanthus	annual tm > 10		Jan tmin > −3	600–1500		5–8	5.5–7.5	Medium	Deep	Fine, Medium – Fine	Coarse, Medium	Low	Low	Non-tolerant	Preference for SW - S - SE orientation. Thermal time > 400 -600° C-day

(continued)

Table 7 (continued)

Target energy crop	Climatic variables					Soil variables								Flooding	Other
	Mean temperature (tm, °C)		Minimum temperature (tmin, °C)	Rainfall (mm)		pH		Soil depth		Texture		Salinity (dS/m)			
	Absolute	Optimal	Absolute	Absolute	Optimal	Absolute	Optimal	Absolute	Optimal	Absolute	Optimal	Absolute	Optimal		
Switchgrass	6 < annual tm < 36; May–Sep > –1	17–32	Jan tm > –10; Oct tm> –1	350–2700	500–1000	4.9–8.2	6–7	Shallow	Medium	Medium – Fine, Medium	Coarse	Medium	Low	Some cultivars are moderately tolerant	
Reed canary grass	2–38	17–25	Jan tmin > –5.3	260–2600	700–1500	3.6–8.2	6–7	Shallow	Medium	Fine, Medium – Fine	Medium, Coarse	Low	Low	(not reported)	Preference for South orientation.
Rye	3–31	15–20	Jan tmin > –8.1; Nov tmin> –1	330–2000	600–1000	3.6–8.2	5.5–6	Shallow	Medium	Fine, Medium – Fine	Medium, Coarse	No limiting	Low	Non-tolerant	
Biomass sorghum	May–Sep (>10°C)	11 < annual tm< 18.6	Jan tmin > –2; May tmin> –1	180–3000	500–1000	3.8–7.7	6–7	Shallow	Medium	Fine, Coarse	Medium, Medium - Fine	Medium	Low	Non-tolerant	Thermal time > 1500 ° -day
Triticale	Jun–Ag.> 13° C		Jan tmin > –12	200–800	400–800	4.5–8	5–7.5	Shallow	Medium	Fine, Coarse	Medium, Medium - Fine	Medium	Low	Non-tolerant	Better tolerance to flooding than wheat
Red gum (E.camaldulensis)	7–40	12–20	Jan tmin > –5	284–2500	600–1000	4.1–7.8	5–7	Medium	Deep	Fine, Medium – Fine	Medium, Coarse	Medium	Low	Moderately tolerant	

Blue gum (E.globulus)	6–35	10–15.5	Jan. tmin > –3	284–1800	700–1500	3.8–7.8	5.5–6.5	Medium	Deep	Medium–Fine	Medium, Coarse	Low	Low	Non-tolerant	
Poplar	8–34	14–30	Jan tmin > –8.5; Oct tmin > -1	235–3000	480–2500	3.6–7.7	6.5–7	Medium	Deep	Fine, Coarse	Medium, Medium-Fine	Medium	Low	Moderately tolerant	
Black locust	6–40	15–32	Jan tmin > –5.7	300–1600	450–700	3.8–7.7	5.5–6.5	Medium	Deep	Coarse	Medium-Fine, Medium	Medium	Low	Non-tolerant	Sensitive to shade; not in NW-N-NE.
Willow	tm March > 5	15–26	Jan tmin > –8.1	310–2000	600–700	3.6–8.5	5.5–7.5	Medium	Deep	Fine, Medium – Fine	Medium	Low	Low	Tolerant	Preference for S orientation
Siberian elm	10.4 < annual tm < 18.5	13.7 <annual tm< 17.2	Jan tmin > –10°C	>300	380–510	4–7.7	–	Medium	Deep	Medium – Fine, Medium	Coarse	Medium	Low	Non-tolerant	Preference for high solar irradiation; SW – S - SE orientation. Thermal time about 400 - 600 ° C.day

many regions of Europe, are high water-demanding; however, they would require irrigation in most arable lands of Spain.

Regarding the dry arable lands of Spain, this study focused on a total of eight plant species. Results of dry arable land suitability for growing those plant species as energy crops are given in Fig. 9. According to our approach, triticale and Siberian elm (SRC) showed the highest potential (=the largest share in category VS). Triticale is well adapted to the dry conditions of Spain. In fact, it has been increasingly grown as a cereal in dry farming in Spain, representing nearly 2% of the field crop land area (cultivated dry land area) in 2013 (MAGRAMA 2015).

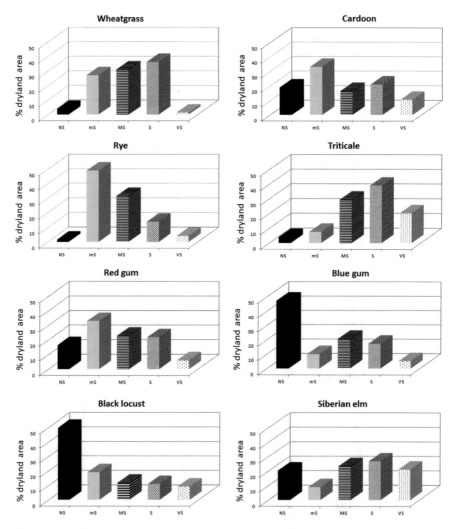

Fig. 9 Land suitability of potential energy crops for dry arable lands in Spain. NS not suitable, mS marginal suitability, MS moderate suitability, S suitable, VS very high suitability

The same source (MAGRAMA 2015) reports that 96,242 t of straw from wheat, barley and oat has been used in 2013 for bioenergy in Spain; in principle, it seems there is not technical reason not to use straw from triticale as well. Contrary to triticale, Siberian elm is woody and little is known about its potential for bioenergy. Adaptive advantages of Siberian elm to harsh environments (Dulamsurena et al. 2009), experiments of coppicing Siberian elm (Pérez et al. 2014; Sanz et al. 2011a; Geyer 1991) and results of land suitability in this work suggest that it could become an interesting energy crop in Spain.

At the other end of the ranking of dry land suitability, the lowest potential (category NS) was recorded for black locust and eucalyptus. For black locust the limiting conditions were the minimum temperature and the soil depth. Deep soils are also preferred by *Eucaliptus* spp. but in this case other variables like salinity, soil texture and hydric requirements (*E.globulus*) weighed more. Nevertheless, our approach was restricted to the suitability of arable lands; other types of lands –e.g. forest areas and pastures- were not the object of this work. It is worth mentioning that ENCE, the most important Spanish company in the production of bioenergy from forest biomass operates with eucalyptus forest biomass. In addition, it should be highlighted that they are committed to R&D & innovation on other management models of eucalyptus, like high density crops (López and Ruiz 2011). From the side of that company (ENCE 2013) there is a growing interest in *Eucalyptus* spp. for bioenergy production in certain regions of Spain.

Concerning the proposal of energy crops for irrigated lands (Fig. 10), giant reed showed the best result with 81.7% share (sum of VS, S, MS categories) in our approach; the highest potential was allocated to South of Spain (Andalusia). These results are consistent with the fact that giant reed is widespread over the Spanish territory; thus, 698 records of naturalized stands of giant reed can be found in Anthos information system of Spain (www.anthos.es, accessed on 8.04.2016). Poplar was second in rank if VS, S, MS categories are taken as a whole. However, it was first in the ranking of VS category with 29.3% irrigated arable land, mainly concentrated in the fertile lowlands of the rivers Guadalquivir, Guadiana and Ebro. Spain, as a member of the International Poplar Commission (http://www.fao.org/forestry/ipc/en/) (FAO's statutory body) has promoted the cultivation of poplars for wood and fiber products since 1947. According to official information (MAGRAMA 2016b), poplars amounted to 134,146 ha forest area in 2015 having been identified three main regions of poplar stands at the Duero, Ebro and Guadalquivir River Basins. *Populus* spp. has also been studied as SRC energy crop in Spain (Sixto et al. 2010). Contrary to *Populus* spp., the suitability of the irrigated arable land in Spain for growing willow was low due to its different agro-ecological requirements. *Populus* spp. and *Salix* spp. (willows) belong to the same botanical family, *Salicaceae;* both of them can be grown in SRC but willow is more suitable for colder climates than poplar (Venendaal et al. 1997).

As shown in Sect. 3.1 the land area identified as potentially available for bioenergy in Spain would be concentrated on the regions (NUTs2) of Castile-La Mancha (9) and Castile-Leon (see Fig. 7, bar codes 9 and 14, respectively) and it would be mainly allocated to dry farming (rainfed fallow land and abandoned

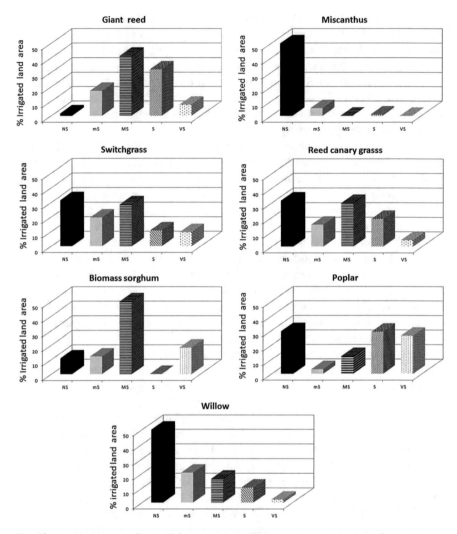

Fig. 10 Land suitability of potential energy crops for irrigated arable lands in Spain. NS not suitable, mS marginal suitability, MS moderate suitability, S suitable, VS very high suitability

farmland) in the two studied scenarios. Main crops in these regions are cereals, representing 57.8% of the cereal land of Spain (↔56.0% of the grain production of Spain). In 2014 the share of wheat and barley was 20.6% and 30.2% respectively, over the cereal area in both regions. On the whole, the suitability of cereal crops to these regions has been proved; nevertheless there is a great diversity among cereals. Concerning lignocellulosic biomass from cereals, the production of cereal straw from Castile-La Mancha and Castile-Leon amounted 4.2 million t in 2014 (↔52.6% of the Spanish straw production) (MAGRAMA 2015); a fraction of that could be harnessed to generate bioenergy in ad-hoc facilities like the straw biomass power plant in Briviesca (Castile-Leon) by ACCIONA (http://www.acciona-energia.com). Within the limits of our suitability assessment, the best crop for the cultivation of fallow and abandoned dry lands in Castile-La Mancha and Castile-Leon would be triticale, for which the categories VS and S amounted to 60%. Siberian elm ranked second with 46%, followed by wheatgrass (40%) (Fig. 11). These crops appear as good candidates for surplus dryland in central Spain. In fact, they have been studied in some locations of Spain in the last years (Ciria et al. 2015b; Fernández et al. 2009; Pérez et al. 2011, 2012, 2014; Sanz et al. 2011a, 2015). To the best of our knowledge, data of yields or performance of these plant species grown for biomass is also scarce in other countries (Geyer et al. 1987; Olovyannikova 1990; Geyer 1991). Further field research would be needed to determine their potential for bioenergy.

4 Conclusions

Under the current circumstances of impacting climate change, a need for the development of renewable energies and the creation of new opportunities for farmers and the recovery of rural areas, the sustainable production of biomass for energy and bio-refineries becomes crucial. In this regard, the present study focuses on the interlinked issues of land availability for dedicated crops and the selection of the species best adapted to the assessed available land in Spain.

According to this study, while the total agricultural area in Spain decreased from 18.5 Mha in 17.1 Mha in the analyzed period (1998–2013), the fallow land, remained to some extent, steady in the same period, amounting to 2,703,260 and 372,821 ha, in rain-fed and irrigated conditions, respectively. The willingness of farmers to leave their land fallow seems to be driven by adaptation strategies to biophysical constraints rather than political measures for food markets regulation (CAP reforms to avoid food surpluses). Thus, proposing alternatives of dedicated crops adapted to limiting growing conditions for turning set-aside agricultural land into production could be a judicious strategy.

Moreover, different constraints as remoteness, low density population, low investment ratios as well as low paid rent conduced in Spain to farmland abandonment. As a result of this study, the abandoned arable land was estimated at 18,164 ha in the case of non-irrigated land and 15,885 ha for irrigated land.

Thus, the agricultural land potentially available for growing dedicated energy crops amounted to 1,572,090 and 3,110,130 ha, depending on the consideration of 50% of fallow land – scenario I (Sc-I)- or the total fallow land – scenario II (Sc-II)- as land for bioenergy purposes.

For the assessed available land in both, rain-fed and irrigated conditions, the agro-ecological suitability for 15 different species of lignocellulosic energy crops was assessed following a multi-criteria decision analysis in a geographical information system (GIS) approach.

For rain-fed conditions, results showed that species like triticale and Siberian elm have the highest potential thanks to their adaptability to harsh and dry conditions. On

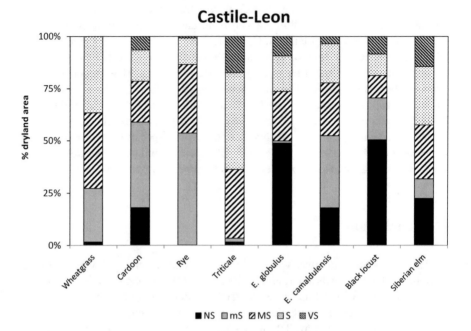

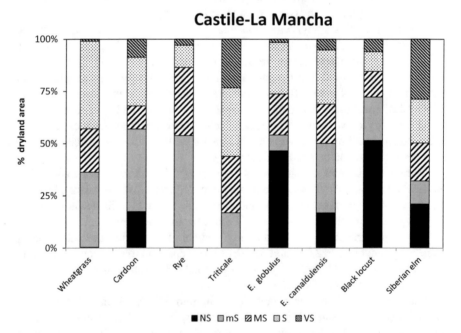

Fig. 11 Land suitability of potential energy crops for dry arable lands in Castile-Leon and Castile-La Mancha. NS not suitable, mS marginal suitability, MS moderate suitability, S suitable, VS very high suitability

the other hand, other species such as black locust showed lower levels of suitability in the conditions of the agricultural areas of Spain (soil depth, water availability).

In the available irrigated lands, giant reed appeared to be the crop with the largest potential followed by poplar, which at the same time are very widespread in the Spanish territory as both, naturally and cultivated stands. Willow on the other hand, showed lower adaptability to the Mediterranean agro-climatic characteristics.

References

AEBIOM (2015) AEBIOM statistical report 2015. European Bioenergy Outlook. European Biomass Association

AEMET (Agencia Estatal de Meteorología-Meteorological Agency of Spain) (2012) Climatic data of Spain. Ministry of Agriculture, Food and Environment of Spain. www.aemet.es

AEMET (Agencia Estatal de Meteorología-Meteorological Agency of Spain) (2014) Climatic monthly summaries 2013–2014. Ministry of Agriculture, Food and Environment of Spain. Available at: http://www.aemet.es/es/serviciosclimaticos/vigilancia_clima/resumenes?w=0&datos=0. Accessed 29 May 2016

ADEME (Agence de l'environment et de la maîtrise de l'energie) (1998) Phalaris. 4 pp.

AGAPA (Agencia de Gestión Agraria y Pesquera de Andalucía) (2012) Ensayos con cultivos energéticos. Periodo 2005–2010. Síntesis de resultados y principales conclusiones. Consejería de Agricultura, Pesca y Medio Ambiente. Junta de Andalucía. 88p

Allen B, Kretschmer B, Baldock, D, Menadue H, Nanni S, Tucker G (2014) Space for energy crops – assessing the potential contribution to Europe's energy future. Report produced for BirdLife Europe, European Environmental Bureau and Transport & Environment. IEEP, London. Available at: http://www.eeb.org/EEB/?LinkServID=F6E6DA60-5056-B741-DBD250D05D441B53. Accessed 16 May 2016

Andersson B, Lindvall E (2002) Development of reed canarygrass varieties for bioenergy production. 12th European on biomass for energy, industry and climate protection, 17–21 June 2002, Amsterdam, The Netherlands. Vol. I pp. 202–203

ANTHOS (2012–2016) Sistema de Información sobre las plantas en España. Fundación Biodiversidad-Ministerio de Agricultura, Alimentación y Medio Ambiente & Real Jardín Botánico – Consejo Superior de Investigaciones Científicas. www.anthos.es

Aylott M, McDermott F (2012) Domestic Energy Crops; Potential and Constraints Review. Project Number: 12–021. NNFCC report. 27 pp. Available at: https://www.gov.uk/government/uploads/system/uploads/attachment_data/file/48342/5138-domestic-energy-crops-potential-and-constraints-r.PDF. Accessed 12 June 2016

Baldelli C (1996) *Robinia* SRF exploitation and conversion: the last researchers. 9th European bioenergy conference. Biomass for energy and the environment, 24–27 Jun 1996, Copenhagen, Denmark, pp. 686–690

Bao M, Rodríguez JL, Crespo I, Lamas J (1996) Miscanthus plantations in Galicia N.W. of Spain. Results and experience over the last three years. Proceedings of the 9th EUropean Conference on Biomass, pp. 608–612

Cañellas I, Huelin P, Hernández MJ, Ciria P, Calvo R, Gea-Izquierdo G, Sixto H (2012) The effect of density on short rotation *Populus* sp. plantations in the Mediterranean area. Biomass Bioenergy 46:645–652

Carrasco J, Del Val MA, Maletta E, Pérez P, Pérez J, Ciria P (2014) Cultivos herbáceos anuales para producción de biocarburantes en España en el marco del proyecto singular estratégico ON-CULTIVOS. Colección documentos CIEMAT. Editorial Ciemat, Madrid, 2014, 111 pp. ISBN: 978–84–7834-727-8. NIPO: 721–14–057-1

Ceulemans R, McDonald AJS, Pereira JS (1996) A comparison among eucalypt, poplar and willow characteristics with particular reference to a coppice, growth-modelling approach. Biomass Bioenergy 11(2/3):215–231

Christou M, Mardikis M, Alexopoulou E (2000) Giant reed *(Arundo donax* L.) network. Improvement, productivity and biomass quality. Proceedings of the 1st World Conference on Biomass for Energy and Industry, Seville, Spain, 5–9 June 2000. ETA-Florence, pp. 1803–1806

Ciria P (2009) El chopo como cultivo energético. Hoja divulgadora n° 2131 HD. MARM

Ciria P, Ciria CS, Maletta E, Pérez J, Barro R, Carrasco J (2015a) Varietal response of *Elytrigia elongata* under semiarid conditions in Spain: Establishment. Proceedings of the 23rd EU Biomass Conference and Exhibition, pp. 139–146. ISBN: 978-88-89407-516. DOI: https://doi.org/10.5071/23rdEUBCE2015-1CV.1.8

Ciria P, Maletta E, del Val MA, Pérez J, Pérez P, Carrasco J (2015b) Cultivos herbáceos anuales para producción de biomasa en España en el marco del Proyecto Singular Estratégico On-cultivos. Colección Documentos Ciemat. ISBN 9788478347322, 116 p

CNE (2013) Informe anual sobre el uso de biocarburantes correspondiente al ejercicio 2011. Comisión Nacional de Energía, 28 de febrero de 2013, 106 pp. Available at: http://energia.cnmc.es/cne/doc/publicaciones/PA012_11.pdf

Cosentino SL, Scordia D, Sanzone E, Testa G, Copani V (2014) Response of giant reed (*Arundo donax* L.) to nitrogen fertilization and soil water availability in semi-arid Mediterranean environment. Eur J Agron 60:22–32

Curt MD (2009a) El cultivo de la caña para producción de biomasa. Hoja Divulgadora núm. 2129 HD. Ministerio de Medio Ambiente y Medio Rural y Marino, I24 pp. ISBN: 978-84-491-0891-4

Curt MD (2009b) Cultivo energético de switchgrass o panizo de pradera (*Panicum virgatum* L.). Hojas Divulgadoras, núm. 2134 HD. Ministerio de Medio Ambiente y Medio Rural y Marino, 24 pp. ISBN: 978-84-491-0893-8

Curt MD, Fernández J, González J, Gil JL (2001) Comparative growth analysis of two sorghum cultivars in Badajoz (Spain). Proceedings of the 1st World Conference on Biomass for Energy and Industry. Vol II, pp. 1877–1880. ISBN: 1-902916-15-8

Curt MD, Mosquera F, Sanz M, Sánchez J, Sánchez G, Esteban B, Fernández J (2012) Effect of land slope on biomass production of *Cynara cardunculus* L. Proceedings of the 20th EU Biomass Conference and Exhibition, pp. 186–190. ISBN 978-88-89407-54-7. doi: https://doi.org/10.5071/20thEUBCE2012-1CO.6.3

Curt MD, Sanz M, Esteban B, Sánchez G, Barreiro M, Fernández J(2011) Assessment of commercial varieties of sorghum as short cycle crops for biomass and sugars production. Proceedings of the 19TH EU biomass Conference and Exhibition, pp. 157–163, ISBN 978-88-89407-55-7. doi: https://doi.org/10.5071/19thEUBCE2011-OC1.3

Dalianis CD, El Bassam N (2010) Giant reed (*Arundo donax* L.) In: Handbook of bioenergy crops, pp. 193–199. Earthscan. ISBN: 978-1-84407-854-7

Dauber J, Brown C, Fernando AF, Finnan J, Krasuska E, Ponitka J, Styles D, Thrän D, Van Groeningen KJ, Weih M, Zah R (2012) Bioenergy from "surplus" lands: environmental and socio-economic implications. BioRisk 7:5–50

Djomo SN, AAc Z, de Groote T, Bergante S, Facciotto G, Sixto H, Ciria P, Weger J, Ceulemans R (2015) Energy performances of intensive and extensive short rotation cropping systems for woody biomass production in the EU. Renew Sust Energ Rev 41:845–854

Dulamsurena C, Haucka M, Nyambayar S, Bader M, Osokhjargal D, Oyungerel S, Leuschner C (2009) Performance of Siberian elm (*Ulmus pumila*) on steppe slopes of the northern Mongolian mountain taiga: drought stress and herbivory in mature trees. Environ Exp Bot 66(2009):18–24. https://doi.org/10.1016/j.envexpbot.2008.12.020

EcoCrop (2013) Ecocrop database (http://ecocrop.fao.org/ecocrop/srv/en/home). FAO, Rome, Italy

EEA (European Environment Agency) (2016) Corine land cover 2012. V.18.5. ESRI Geodatabase. Available at: http://land.copernicus.eu/pan-european/corine-land-cover

El Bassam N (2010) Handbook of bioenergy crops. Earthscan. 516 pp. ISBN: 978-1-84407-854-7

Elbersen HW (Coordinator) (2002). Switchgrass (*Panicum virgatum* L.) as an alternative energy crop in Europe. Initiation of a productivity network. Final Report for the period 1/04/1998-30/09/2001. EU Project FAIR 5-CT97–3701

ENCE (2013) Activity Report. Available on the internet at: https://www.ence.es/images/pdf/memoria%20ence%202013_BAJA%20ING-0807943.pdf

Estel S, Kuemmerle T, Alcántara C, Levers C, Prishchepov A, Hostert P (2015) Mapping farmland abandonment and recultivation across Europe using MODIS NDVI time series. Remote Sens Environ 163:312–325

European Commission (2010) Rural Development in the European Union. Statistical and Economic Information. Report 2010. DG AGRI, Brussels

European Parliament (2009) Directive 2009/28/EC of the European Parliament and of the Council of 23 April 2009 on the promotion of the use of energy from renewable sources and amending and subsequently repealing Directives 2001/77/EC and 2003/30/EC. Off J Eur Union. 5 June 2009

European Parliament (2013) EU 1307/2013: Regulation (EU) No 1307/2013 of the European Parliament and of the Council of 17 December 2013 establishing rules for direct payments to farmers under support schemes within the framework of the common agricultural policy and repealing. Counc Off J Eur Union, pp. 608–670

European Parliament (2015) Directive 2015/1513 of the European Parliament and of the Council of 9 September 2015 amending Directive 98/70/EC relating to the quality of petrol and diesel fuels and amending Directive 2009/28/EC on the promotion of the use of energy from renewable sources. Off J Eur Union. 15 Sept 2015

EUROSTAT - Statistical Office of the European Union (2015) Land cover, land use and landscape. Data extracted in March 2015. http://ec.europa.eu/eurostat/web/lucas/publications. Accessed on 13 April 2016

Facciotto G, Di Candilo M, Bergante S, Baratto G, Diozzi M (2008) Willow clones' for biomass production in SRC plantations. 16th European Biomass conference & Exhibition, pp. 611–613

FAO (1996) Agro-ecological zoning. Guidelines. FAO Soils Bulletin. p. 73

FAO (2014) FAOSTAT, Methods & standards. http://faostat3.fao.org/mes/glossary/E. Accessed 16 May 2016

FAO/IIASA/ISRIC/ISS-CAS/JRC (2008) Harmonized World Soil Database (version 1.0). FAO, Rome, Italy and IIASA, Laxenburg, Austria. http://www.fao.org/fileadmin/templates/nr/documents/HWSD/HWSD_Documentation.pdf

Fernández J (2009) El cultivo del cardo para producción de biomasa. Hoja divulgadora n° 2130 HD. MARM (2009)

Fernández J, Curt MD, Aguado PL (2006) Industrial applications of *Cynara cardunculus* L. for energy and other uses. Ind Crop Prod 24(3):222–229

Fernández J, Iriarte L, Sanz M, Curt MD (2009). Preliminary study of Siberian elm as an energy crop in a continental-Mediterranean climate. Proceedings of the 17TH EU Biomass Conference and Exhibition, pp. 148–153. ISBN 978-88-89407-57-3

Geyer WA (1991) Coppice response to cut stump height of three hardwood tree species. In: Grassi G, Colina A, Zibetta H (eds) Proceedings of the 6th European Conference on Biomass for Energy, Industry and Environment. 21–27 Abril, Atenas (Grecia), pp. 207–211

Geyer WA, Argent RM, Walanwender WP (1987) Biomass properties and gasification behaviour of 7-year-old Siberian elm. Wood Fiber Sci 19(2):176–182

Gominho J, Lourenco A, Palma P, Lourenco ME, Curt MD, Fernández J, Pereira H (2011) Large scale cultivation of *Cynara cardunculus* L. for biomass production. A case study. Ind Crop Prod 33:1–6

Habayarimana E, Bonardi P, Laureti D, Di Bari V, Cosentino S, Lorenzoni C (2004) Multilocal evaluation of biomass sorghum hybrids under two stand densitites and variable water supply in Italy. Ind Crop Prod 20(1):3–9

Hart K, Allen B, Lindner M, Keenleyside C, Burgess P, Eggers J, Buckwell A (2013) Land as an Environmental Resource, Report Prepared for DG Environment, Contract No ENV.B.1/ETU/2011/0029, Institute for European Environmental Policy, London

Hastings A, Clifton-Brown JC, Wattenbach M, Mitchell CP, Stampfl P, Smith P (2009) Future energy potential of Miscanthus in Europe. Glob Change Biol Bioenergy 1:180–196

IIASA/FAO (2012) Global Agro-ecological Zones (GAEZ v3.0). IIASA, Laxenburg, Austria and FAO, Rome, Italy

Iriarte L (2008) Caracterización del olmo de Siberia (*Ulmus pumila* L.) como cultivo energético. PhD Thesis. Technical University of Madrid, Spain

JRC -Joint Research Centre (2010) Map of Soil pH in Europe. Land Management and Natural Hazards Unit, Institute for Environment & Sustainability, European Commission

Keenleyside C, Tucker GM (2010) Farmland abandonment in the EU: an assessment of trends and prospects. Report prepared for WWF. Institute for European Environmental Policy, London. Available at: http://www.ieep.eu/assets/733/Farmland_abandonment_in_the_EU_-_assessment_of_trends_and_prospects_-_FINAL_15-11-2010_.pdf. Accessed 02 June 2016

Lavalle C, Baranzelli C, Mubareka S, Rocha C, Hiederer R, Batista F, Estreguil C (2011). Implementation of the CAP Policy Options with the Land Use Modelling Platform. A first indicator-based analysis. Institute for Environment & Sustainability, European Commission DG Joint Research Centre. Scientific and Technical Reports. EUR 24909. 130 pp. ISBN 978-92-79-20917-8. doi: https://doi.org/10.2788/45131

Lewandowski I, Scurlock JMO, Lindvall E, Christou M (2003) The development and current status of perennial rhizomatous grasses as energy crops in the US and Europe. Biomass Bioenergy 25:335–361

López G, Ruiz F (2011) Cultivos energéticos y biomasa forestal. ENCE. Jornadas Bosques en Cambio: pasado y futuro. Huelva, 24 octubre 2011

MAGRAMA (Ministerio de Agricultura, Alimentación y Medio Ambiente) - *Ministry of Agriculture, Food and Environment of Spain* (2015) Anuario de Estadística- Annual Agricultural Statistics (AAS) 1998–2013. España. Available at: http://www.magrama.gob.es/es/estadistica/temas/publicaciones/anuario-de-estadistica/. Accessed 03 Feb 2016

MAGRAMA (Ministerio de Agricultura, Alimentación y Medio Ambiente) - *Ministry of Agriculture, Food and Environment of Spain* (2016a) Libro Digital del Agua: Servicios Hídricos, Precipitación. Available at: http://servicios2.marm.es/sia/visualizacion/lda/recursos/climatologia_precipitacion.jsp. Accessed 20 July 2016

MAGRAMA (Ministerio de Agricultura, Alimentación y Medio Ambiente) (2016b) Encuesta sobre superficies y rendimientos de cultivos. ESYRCE. N.I.P.O.: 280-15-015-2

Maletta E, Carrasco J, Del Val MA, Pérez J, Ciria CS, Ciria P (2016) Cultivos herbáceos perennes para la producción de biomasa en España. Colección Documentos Ciemat

McLaughlin SB, Kszos LA (2005) Development of swithgrass (*Panicum virgatum*) as a bioenergy feedstock in the United States. Biomass Bioenergy 28:515–535

Meehan TD, Gratton C, Diehl E, Hunt ND, Mooney DF, Ventura SJ et al (2013) Ecosystem-service tradeoffs associated with switching from annual to perennial energy crops in riparian zones of the US Midwest. PLoS One 8(11):e80093. https://doi.org/10.1371/journal.pone.0080093

Montoya JM (1988) Chopos y choperas. Ed. Mundi-Prensa. Madrid

Montoya JM (1995) El Eucalipto. Ed. Mundi-Prensa. 125 pp.

Montoya JM, Mesón M (2004) Selvicultura. Ed. Mundi-Prensa. Tomo I

Olovyannikova IN (1990) Growth or the coppice regeneration of Siberian elm on soils of the alkali complex in the Caspian region. Sov For Sci 3:18–26

Panagos P (2006) The European soil database. GEO: connexion 5(7):32–33

Panagos P, Van Liedekerke M, Jones A, Montanarella L (2012) European soil data centre: response to European policy support and public data requirements. Land Use Policy 29(2):329–338. https://doi.org/10.1016/j.landusepol.2011.07.003

Pereira H, Pereira JS, Madeira M, Araujo C, Almeida H, Leal LC, Chaves M, Fabiao A (1990). Optimization of biomass production in short rotation *Eucalyptus globulus* plantations. Commission of the European Communities. Final Report. EUR 12718 EN (1990), 34 pp.

Pérez I (2016) Evaluación de *Ulmus pumila* y *Populus* spp. como cultivos energéticos en corta rotación. PhD Thesis. Technical University of Madrid, Spain

Pérez I, Ciria P, Bergante S, Pérez J, Carrasco J, Rosso L, Facciotto G (2012) Biomass production of Siberian elm at the end of the second vegetative period in Spain and Italy. Proceedings of the 20th EU Biomass Conference and Exhibition, pp. 402–408. ISBN: 978-88-89407-54-7. DOI: https://doi.org/10.5071/20thEUBCE2012-1CV.3.8

Pérez I, Pérez J, Carrasco J, Ciria P (2011) *Ulmus pumila* under irrigated and rain-fed continental conditions in the first year of growth. Proceedings of the 19th EU Biomass Conference and Exhibition, pp. 765–769. ISBN: 978-88-89407-55-7. doi: https://doi.org/10.5071/19thEUBCE2011-VP1.3.75

Pérez I, Pérez J, Carrasco J, Ciria P (2014) Siberian elm responses to different culture conditions under short rotation forestry in Mediterranean areas. Turk J Agric For 38:652–662

Piquero N, Curt MD, Fernández J (2011) Performance of five close relatives of wheatgrass grown for biomass in a small scale experiment. Proceedings of the 19TH EU Biomass Conference, pp. 623–628. ISBN 978-88-89407-55-7. doi: https://doi.org/10.5071/19thEUBCE2011-VP1.3.26

Qingguo X, El Bassam N (2010) Miscanthus (*Miscanthus* spp.). In: Handbook of bioenergy crops, pp. 240–251. Earthscan. ISBN: 978-1-84407-854-7

Raggi M, Sardonini L, Viaggi D (2013) The effects of the common agricultural policy on exit strategies and land re-allocation. Land Use Policy 31:114–125. https://doi.org/10.1016/j.landusepol.2011.12.009

Renwick A, Jansson T, Verburg PH, Revoredo-Giha C, Britz W, Gocht A, McCracken D (2013) Policy reform and agricultural land abandonment in the EU. Land Use Policy 30:446–457. https://doi.org/10.1016/j.landusepol.2012.04.005

Robledo A, Correal E (2013). Cultivos energéticos de segunda generación para producción de biomasa lignocelulósica en tierras de cultivo marginales. Murcia: Instituto Murciano de Investigación y Desarrollo Agrario y Alimentario, 159 p. Available at: www.imida.es

Sánchez J, Curt MD, Fernández J (2016) Approach to the potential production of giant reed in surplus saline lands of Spain. Glob Change Biol Bioenergy., 14 pp 9:105. https://doi.org/10.1111/gcbb.12329

Sanz M, Curt MD, Plaza A, García-Müller M, Fernández J (2011a) Assessment of Siberian elm coppicing cycle. Proceedings of the 19th EU Biomass Conference and Exhibition, pp. 601–605, ISBN 978-88-89407-55-7. doi: https://doi.org/10.5071/19thEUBCE2011-VP1.3.18

Sanz M, Mosquera F, Sánchez G, Barreiro M, Aguado PL, Sánchez J, Curt MD, Fernández J (2011b). Biomass potential of triticale and rye in two different phenological stages and energy characterization. Proceedings of the 19th EU Biomass Conference and Exhibition, pp. 591–595, ISBN 978-88-89407-55-7. doi:https://doi.org/10.5071/19thEUBCE2011-VP1.3.16

Sanz M, Sánchez J, Mosquera F, Curt MD, Fernández J (2015) Influence of planting season on Siberian elm yield and economic prospects. Proceedings of the 23rd EU Biomass Conference and Exhibition, pp. 198–204. 978-88-89407-516. doi: https://doi.org/10.5071/23rdEUBCE2015-1CV.1.38

Sauer J, Davidova S, Latruffe L (2012) Determinants of smallholders' decisions to leave land fallow: the case of Kosovo. J Agric Econ 63:119–141. https://doi.org/10.1111/j.1477-9552.2011.00321.x

Sixto H, Hernández MJ, Ciria MP, Carrasco J, Cañellas I (2010) Manual de cultivo de *Populus* spp. para la producción de biomasa con fines energéticos. INIA. ISBN: 978-84-7498-530-6, 57 pp.

Sixto H, Hernández MJ, de Miguel J, Cañellas I (2013) Red de parcelas de cultivos leñosos en alta densidad y turno corto. INIA. ISBN 978-84-7498-559-7, 31 pp.

Terres JM, Nisini L, Anguiano E (2013) Assessing the risk of farmland abandonment in the EU. Institute for Environment & Sustainability, European Commission DG Joint Research Centre. Scientific and Policy Reports. EUR 25783 EN, 132 pp. ISBN 978-92-79-28281-2. doi: https://doi.org/10.2788/81337

Toth G (2008) Map of Saline and Sodic Soils in the EU. Land Management and Natural Hazards Unit, Institute for Environment & Sustainability, European Commission DG Joint Research Centre

USDA, NRCS (2016) The Plants Database (http://plants.usda.gov, 19 March 2016). National Plant Data Team, Greensboro, NC 27401-4901 USA

Utrilla V, Sturzenbaum MV, Rivera EH (2007) Evaluación de agropiros de origen canadiense y nacional en EA. Punta Loyola. INTA. EEA Santa Cruz. Macro región Patagonia sur. AER Río Gallegos. Argentina

Van Liedekerke M, Jones A, Panagos P (2006) ESDBv2 Raster Library – a set of rasters derived from the European Soil Database distribution v2.0 (published by the European Commission and the European Soil Bureau Network, CD-ROM, EUR 19945 EN)

Venendaal R, Jorgensen U, Fosters CA (1997) European energy crops: a synthesis. Biomass and Bioenergy 13(3):147–185

Walsh M, Jones MB (2000) Miscanthus for energy and fibre. Pub. Earthscan. ISBN 9781902916071, 202 p

WBA (World Bioenergy Association) (2014) WBA Global Bioenergy Statistics 2014, 39 pp.

Welfle A, Gilbert P, Thornley P (2014a) Securing a bioenergy future without imports. Energy Policy 68:1–14

Welfle A, Gilbert P, Thornley P (2014b) Increasing biomass resource availability through supply chain analysis. Biomass Bioenergy 70:249–266

Areas with Natural Constraints to Agriculture: Possibilities and Limitations for The Cultivation of Switchgrass (*Panicum Virgatum* L.) and Giant Reed (*Arundo Donax* L.) in Europe

Parenti Andrea, Lambertini Carla, and Andrea Monti

Abstract The European Union is facing complex issues concerning the achievement of GHG reduction goals. CO_2 emissions could be reduced by a rational allocation of perennial lignocellulosic crops on unsuitable land for agriculture. These crops would also potentially reverse the increasing trend of soil depletion and land abandonment in Europe. The Joint Research Centre (JRC) identified biophysical constraints aimed at defining and unifying the definition of marginal land across Europe, and evaluating agricultural opportunities in these areas. In this study we evaluate possibilities and limitations for the cultivation of two of the most promising perennial biofuel crops (giant reed and switchgrass) in Europe, in areas with natural constraints (ANC land), as identified by the JRC. Based on the literature, both giant reed and switchgrass appear suitable for ANC land. Only shallow rooting depth and waterlogging can limit the establishment and agricultural mechanization of these rhizomatous plant species. Field tests in ANC land are needed to assess the potential yields provided by lignocellulosic crops under such limiting conditions. These results are fundamental to stimulate farmers' acreage expansion and the development of a supply chain. Further research in the impact of lignocellulosic crops on the evolution of ANC ecosystems and lignocellulosic species is also needed to ensure a sustainable use of ANC land.

Keywords Biophysical constraints · Marginal land · Land suitability · Giant reed · Switchgrass

P. Andrea · L. Carla
Department of Agricultural and Food Sciences, University of Bologna, Bologna, Italy

A. Monti (✉)
Alma Mater Studiorum – University of Bologna, Bologna, Italy
e-mail: a.monti@unibo.it

© Springer International Publishing AG, part of Springer Nature 2018   39
R. Li, A. Monti (eds.), *Land Allocation for Biomass Crops*,
https://doi.org/10.1007/978-3-319-74536-7_3

1 Introduction

The European Union (EU) is facing complex issues concerning the reduction of greenhouse gasses (GHG) emissions, enhancement of renewable energy and land abandonment. To cut the 36.9% of the GHG emissions in the atmosphere by 2030 (from 2005 base level) is an ambitious target that needs concrete actions in order to slow down the global warming process. The pathway that has been undertaken concerns the transition towards renewable energies alongside the fossils feedstock, which is still predominant with a 78% of the total primary energy consumption in the 28 countries of the European Union (EU-28) (European Environmental Agency 2008). Dedicated non-food lignocellulosic biomass crops represent an important source of renewable energy and are regarded by the European Commission as a viable energy feedstock to fulfill the pressing targets. This transition is mainly hampered by the competition for land with food and feed crops, and by a limited development of sustainable supply chains of lignocellulosic biomass. Not last, farmers' scepticism for long term investment on unknown crops is also holding lignocellulosic agriculture from taking off.

Land abandonment is an increasing phenomenon in many European regions. Climate change and decreasing soil fertility due to drought and desertification are threatening Mediterranean agriculture and reducing land surfaces suitable for food production. In many mountain and hill areas agriculture is no longer economically feasible due to low yields and mechanization difficulties. Small farm size, farmers' old age, weak land market, low population density and remoteness are all factors that are accelerating land abandonment.

Between 1961 and 1999 the total Utilized Agricultural Area (UAA) including arable land, permanent pastures, meadows and permanent crops, decreased by 19 million hectares (Mha) in EU-27 (FAOSTAT 2012). From 2000 to 2006 the UAA decreased further by 3 Mha (European Environmental Agency 2007), and in 2010 it was estimated that 13.2 Mha (mainly fallow land), accounting for 7.25% of the EU-27 total UAA, were no longer used for food/feed production (Krasuska et al. 2010).

The report of the Joint Research Centre (JRC) (Terres et al. 2013) 'Assessing the risk of farmland abandonment in the EU' predicts that in the next 20 and 30 years the trend of land abandonment will likely continue in Europe due to high competition from the global market and decreasing support for extensive farming from the Common Agricultural Policy (CAP). Likewise, the study of Krasuska et al. (2010) shows that arable land not suitable for food production will reach 20.5 Mha in 2020 and 26.3 Mha in 2030.

The increase of unmanaged abandoned land raises concerns about water runoff, erosion, loss of soil fertility, water logging, salinization, sodification and desertification which poes a threat to agriculture recovery, but also affects natural succession, and can lead to land degradation (Bonet 2004). An option to mitigate these problems is given by the cultivation of permanent lignocellulosic crops and the creation of innovative logistics concepts in abandoned UAA which can have positive economic, environmental and social outcomes.

2 Area with Natural Constraints: Policy Context and Background

The updated classification of European marginal land is based on the former 'Less Favoured Areas' (LFA) classification and aims at localizing land with natural constraints to agriculture and prevent rural depopulation and land abandonment (Jones et al. 2014). Three categories underpinned the LFA classification:

- **Mountain areas:** the limiting factors are steep slopes and a short growing season due the harsh climate of the high altitudes. Latitudes over the 62nd parallel are characterized by a short growing season and are also included in this category.
- **Intermediate or other than mountain areas:** are areas with natural handicaps to agriculture like poor climate conditions or low soil productivity. Management of these lands through an extensive farming is important to prevent land degradation.
- **Areas affected by specific handicaps:** are areas which need management in order to avoid economic and environmental depletion (coastline protection, countryside and tourism conservation).

With the exception of the 'Mountain areas', that maintained the same name, the regulation of the European Commission 1305/2013, Art. 32.4 updated the LFA by changing the denomination and delimitation of 'Intermediate or other than mountain areas' and 'Areas affected by specific handicaps' into 'Areas facing natural constraints' and 'Areas with specific constraints' respectively. The LFA system was so replaced by the current classification in 'Area with natural constraints' (ANC) (also adopted in this study), and which is based on the biophysical criteria defined by JRC. The Rural Development Regulations for the period 2014–2020 introduced the obligation for the EU Member States (MS) to map areas with such natural constrains in order to extend CAP support. This allowed the detection of unused land with higher accuracy and the design of programmed and safer management measures meant to achieve environmental, economic and social benefits (Confalonieri et al. 2014).

3 Biophysical Constraints to Define ANC Land

The biophysical parameters identified as constraints for agriculture and used by the European Commission to classify ANC land concern climate, soil and land suitability for agriculture. The criteria and thresholds defined by the JRC experts

(Table 1) are derived from an elaboration of the Food and Agriculture Organization's (FAO) approach concerning degraded land, and the agronomic Liebig's law (Jones et al. 2014). FAO considered the mutual interactions of biophysical land characteristics and overall land suitability to agriculture for a wide range of crops (FAO and IIASA 2000), whereas Liebig's law of the minimum states that plants' growth is not controlled by the total amount of available resources, but by the scarcest one (i.e. the limiting factor). The definition of the biophysical constraints followed by JRC considers also the limits for mechanization according to the conventional mechanized European farm unit producing grain crops or grass (Confalonieri et al. 2014).

3.1 Criteria

The eight main biophysical constraints for agriculture identified by JRC (Table 1) (Jones et al. 2014) are:

- **Low temperature:** in terms of Length of Growing Period (LGP) defined by number of days with daily average temperature > 5 °C (LGP_{t5}) or Thermal-time sum (TS_5) (degree-days) for growing period, defined by accumulated daily average temperature > 5 °C.
- **Dryness:** measured as the ratio of the annual precipitation (P) to the annual potential evapotranspiration (PET) detected by the United Nations Environment Programme (UNEP) aridity index (AI).
- **Excess of soil moisture:** as number of days at or above field capacity.
- **Limited soil drainage:** defined as areas which are water logged for significant duration of the year.
- **Unfavourable texture and stoniness:** as relative abundance of clay, silt, sand, organic matter in terms of weight and volumetric percentage, considering the coarse material fractions.
- **Shallow rooting depth:** as the depth from soil surface to coherent hard rock or hard pan in centimetres.
- **Poor chemical properties:** as the presence of salts, exchangeable sodium and excessive acidity.
- **Steep slope:** defined as change of elevation with respect to planimetric distance (%).

'Unfavourable texture and stoniness' and 'poor chemical properties' represent broad ranges of different conditions, thus they were divided into additional seven sub-criteria: 'stoniness', 'sand, loamy sand', 'heavy clay', 'organic soil', 'salinity', 'sodicity', and 'pH' (Jones et al. 2014). Each constraint was identified by two thresholds, a severe and a sub-severe one. Severe thresholds are numeric lower and upper limit values of critical factors beyond which the land is not suitable for

Table 1 Biophysical constraints criteria, relative threshold and plant suitability

Biophysical constraints	Description	Severe threshold	Sub-severe threshold	Giant Reed	Switchgrass
Low T	LGP$_5$ (days)	≤180	≤195	Medium suitable	Suitable
	TS$_5$ (C°)	≤1500	≤1575	Medium suitable	Suitable
Dryness	P/PET	≤0.5	≤0.6	Medium suitable	Medium suitable
Excess soil moisture	Number of days at or above field capacity (days)	≥230	≥210	Medium suitable	Medium suitable
Limitate soil drainage	Area which are waterlogged for a significant duration of the year	Wet 80 cm > 6 months, or 40 cm > 11 months. Poorly or very poorly drained. Gleyic colour pattern within 40 cm	No change	Unsuitable	Unsuitable
Unfavourable texture and stoniness	Relative abundance of clay, silt, sand, organic matter (weight %) and coarse material (volumetric %) fractions	≥ 15% of topsoil volume is coarse material, rock outcrop, boulder	≥10% of topsoil volume is coarse material, rock outcrop, boulder	Suitable	Suitable
		Texture class in half or more (cumulatively) of the 100 cm soil surface is sand, loamy sand	Sand, loamy sand in 40% or more within 100 cm surface layer	Very suitable	Very suitable
		Topsoil texture class is heavy clay (≥ 60% clay)	Topsoil texture ≥50% clay	Medium suitable	Low suitability
		Organic soil (organic matter ≥30%) of at least 40 cm	Organic matter ≥30%, of at least 30 cm within 100 cm surface layer	NO DATA	NO DATA
		Topsoil contains 30% or more clay and there are vertic properties within 100 cm of the soil surface	No change	Medium suitable	Low suitability
Shallow rooting depth	Rooting depth (cm)	≤30	≤35	Unsuitable	Unsuitable
Poor chemical properties	Salinity (dS/m)	≥4 dS/m in topsoil	≥3.2 dS/m in topsoil	Suitable	Medium suitable
	Sodicity	≥6 ESP in half or more of the 100 cm layer	≥4.8 ESP in half or more of the 100 cm layer	Medium suitable	Medium suitable
	Topsoil acidity (pH)	≤5	≤5.5	Suitable	Medium suitable
Steep slope	Slope (%)	≥15	≥12	Medium suitable	Suitable

agriculture (Table 1). Sub-severe thresholds are generally 20% (with exceptions) below the severity thresholds and can identify land in which agriculture may be feasible but not very profitable. The combination of sub-severe biophysical constraints can in some cases have a positive interaction and result in land in which agriculture is feasible and profitable (Confalonieri et al. 2014).

The 'Areas facing natural constraints' are defined by one of the eight aforementioned criteria within the range between the sub-severe and severe threshold, while the 'Areas with specific constraints' by the combination of two of the biophysical criteria, without any priority order among constrains, within a margin of 20% of the severe threshold value initially defined (Table 1).

The need of simple rules defining this new ANC land brought to the decision that only pairwise combinations of biophysical criteria where elegible for CAP by the MS (Rural Development Regulation for the period 2014–2020), excluding in this way combinations of three or more criteria.

3.2 Synergies Resulting from Pair-Wise Combinations of Biophysical Constraints

The pairwise combinations of the biophysical criteria defining the aforementioned 'Areas with specific constraints' output six different results: 'not occurring', 'not accepted', 'no interaction', 'negative synergy', 'unclear synergy', and 'positive synergy', accounting for 91 overall combinations.

Nineteen pairwise combinations were found unlikely to occur at the same location (termed 'not occurring'); 21 combinations were considered not possible because the sub-severe threshold could not be defined (or was scientifically questionable) for one of the criteria (not accepted); for 18 combinations, no interaction between the two criteria was found (or interaction was already embedded in the concept of the criterion) (no interaction). Otherwise, synergies resulting from co-occurring biophysical constrains were considered and analysed (negative, unclear and positive synergies).

A 'negative synergy' was defined as the combination of two biophysical sub-severe constraints which results in worsening conditions that lead to circumstances considered too severe for any agricultural activity (e.g. Dryness x Stoniness, the influence of each factor increases the severity of the other). They account for 25 combinations.

An 'unclear synergy' was defined as two sub-severe biophysical constraints that are not related to each other (i.e. Excessive soil moisture x Stoniness). The combined effect, either positive or negative, depends on the specific situation and is unpredictable. The MS have the right to assess case by case and treat the two factors independently. They account for five combinations.

A 'positive synergy' was defined as the combination of two biophysical sub-severe constraints which leads to an improvement of the agronomic condition due to the positive effect that one constraint has on the other (i.e. Stoniness x Organic texture). Positive synergies, accounting for three combinations, drop any severe limitation to agriculture and are no more considered constrains of ANC land (Confalonieri et al. 2014).

4 Rationale for ANC Land Allocation to Giant Reed and Switchgrass

Perennial fast growing crops are considered the best candidates for a sustainable cultivation of ANC land. Annual crops require annual mechanical and chemical operations, in addition to annual sowing, and largely exceed the low input requirements to be competitive in the bio-economy.

Giant reed (*Arundo donax*) and switchgrass (*Panicum virgatum*) are two fast growing perennial grasses with high ecological amplitude (Lewandowski et al. 2003; Perdue 1958), and are already cultivated in many world's regions as ligno-cellulosic crops for bioenergy (Wullschleger et al. 2010; Zegada and Monti 2011, Zegada et al. 2013). Historically, giant reed has been cultivated in Europe and switchgrass in North America, but recently both species have been considered for cultivation in Europe. Giant reed and switchgrass can acclimate to different pedoclimatic conditions and tolerate most of the specific biophysical conditions constraining agriculture in ANC land across the European territory. Both plants are highly tolerant to several of the above mentioned biophysical constraints and can provide a considerable amount of biomass under unfavourable conditions. Switchgrass can be planted and harvested with existing farm machinery and has been extensively studied (Boxes 1 and 2).

Box 1 Main eco-physiological characteristics and biomass yields of giant reed

	Giant reed
Taxonomy	*Arundo donax*, *Poaceae* family.
Distribution range	Native to Asia. Currently spread all over the world from warm-temperate to sub-tropical areas as an invasive species (Lewandowski et al. 2003). In Europe it is found in the warm regions of the Mediterranean basin.
Morphology and phenology	Up to 8–9 m height in dense clumps (Perdue 1958). It develops compact masses of knotty rhizomes. Stems emerge at different times. Late shoots do not grow properly due to shady conditions, and they often fail to grow or even die (Sharma et al. 1998). The inflorescence is an erect terminal panicle with silky hairs. The spikelets contain 2–7 flowers, all bisexual except the reduced uppermost one (Lewandowski et al. 2003).
Propagation	Rhizomes, because seeds are sterile (Mariani et al. 2010). Micropropagated plants (Lewandowski et al. 2003).
Physiology	C3 photosynthetic pathway.
Biomass	Mean aboveground biomass variable from 15 to 35 megagrams of dry weight per hectares per year (mg d.w. $Ha^{-1} y^{-1}$) than 17 mg d.w. Ha^{-1} for 9 consecutive years (Monti and Zegada 2016). Belowground biomass: Reaches 13.6 mg d.w. Ha^{-1} (Monti and Zatta 2009).

   Although not native to Europe (Hardion et al. 2014), giant reed has a long history in the Mediterranean region. The botanical records date back to the Roman civilization (Hardion et al. 2014). It has a broad distribution in the Mediterranean region, but outpost populations are known to occur in Germany (Bacher and Sauerbeck 2001) and England, and are nurtured as ornamental plants in Denmark in gardens and parks (Lambertini C., personal observation). This species has high tolerance to salinity, low pH, pests and flooding (Zegada-Lizarazu et al. 2010). Unlike switchgrass, it is a sterile plant propagated by rhizomes or cuttings (Mariani et al. 2010).

Box 2 Main eco-physiological characteristics and biomass yields of switchgrass

	Switchgrass
Taxonomy	*Panicum virgatum, Poaceae* family, *Panicoideae* subfamily.
Distribution range	North American continent.
Morphology and phenology	Up to 3 m height and reaches a rooting depth of 3.5 m (Weaver 1968). Two spontaneously evolved ecotypes: Upland (most frequently octoploid) is shorter, fine-stemmed, better adapted to dry habitats and has lower resistance to rusts (Moser and Vogel 1995). The lowland ecotype (tetraploid) has coarser stems, longer and more bluish-green leaves, longer ligules, a more bunchy-growth form, is taller, more yielding and is adapted to wet sites. The lowland ecotype reaches maturity in a longer growth period (Moser and Vogel 1995; Sanderson et al. 1996). The inflorescence is an apical panicle which can be from 15 to 50 cm long, and produces fertile seeds.
Propagation	Seeds and rhizomes.
Physiology	C4 photosynthetic pathway.
Biomass	Mean aboveground biomass from 10 mg d.w. $Ha^{-1} y^{-1}$ (Alexopoulou et al. 2015) to 13.8 mg d.w. $Ha^{-1} y^{-1}$ in a 5-year experiment (Nocentini et al. 2015). Belowground biomass: Reaches 8.5 mg d.w. Ha^{-1} (Monti and Zatta 2009).

Switchgrass is a neophyte crop species to Europe and is regarded as a promising energy crop in cooler and more continental regions than those inhabited by giant reed, like the grasslands (steppes) in east-central Europe. It shares several tolerance traits with giant reed, but has also some advantages compared to it, such as better acclimation to short growing seasons, the possibility of choosing the best variety or cultivar for specific biophysical conditions (it is a domesticated species), and lower establishment cost (especially due to the possibility of establishing crops by seed, and use traditional cereal seed drills or hydro-seeding). It is a C4 plant which has evolved CO_2 concentration mechanisms that allow the plant to keep stomata closed during the driest hours of the day in the vegetative season, and limit evapotranspiration, while keeping carbon fixation and thus increasing water use efficiency.

5 Suitability of Giant Reed and Switchgrass to The Specific Biophysical Constraints of ANC Land

We scored the suitability of giant reed and switchgrass to the above described constraints with five degrees of suitability: 'unsuitable', 'low suitable', 'medium suitable', 'suitable', and 'very suitable'. In addition to the ecological suitability of the two species, that was considered for a rationale allocation of these crops, we evaluated other technical factors associated with mechanization, such as trafficability, workability and rootability, which may be influenced by some of the constraints (i.e. excess soil moisture and steep slope) (Table 1).

5.1 Low Temperature

This constraint is calculated by the LGP_{t5}, i.e. number of days with daily average temperatures above 5 °C, or as TS_5 (Thermal Sum) and has a sub-severe to severe threshold of 195 to 180 LGP_{t5} and 1575 to 1550 TS_5.

Giant reed has a broad climatic acclimation capacity and it can grow as far north as Denmark (Lambertini C., personal observation), Germany and the United Kingdom (El Bassam 2010). During the dormant winter phase, killing frost occurs with -5 °C temperatures for more than 5 days, while during the spring this threshold rises up to 0 °C for more than 2 days (El Bassam 2010). Long term field experiments showed a LGP_{t5} amount of 206 (Triana et al. 2014) and 223 (Angelini et al. 2009) days in Central Italy, and a TS_5 requirement of 3000 °C (Nassi o Di Nasso et al. 2011) and a mean of 1843 °C (Angelini et al. 2009) in Pisa coastal plain. These data collected in the Mediterranean region are below, or close to, the sub-severe threshold. It can therefore be expected that giant reed can tolerate lower LGP_{t5} and TS_5, and thus theoretically grow at higher latitudes, however no data are available so far.

Switchgrass is a photoperiod sensitive plant. Flowering requires more than 14 h of darkness per day and this photoperiod limits the distribution range of the species towards north (Sanderson et al. 2012). Long photoperiods cause a longer vegetative growth during the fall and, as a consequence, the plant does not harden properly to survive the first winter frosts (Lewandowski et al. 2003). According to Hsu et al. (1985) the minimum soil temperature for seedling germination of Alamo (lowland ecotype) is 10.3 °C, and temperatures from 15.5 °C to 29.5 °C can speed up growth by 3 days (Hsu et al. 1985; Wolf and Fiske 1995). Open field trials in Ontario have shown that certain switchgrass cultivars can tolerate −22 °C during the dormant phase and have a 50% chance of survival (hence killing frost limit for the species has been set at −20 °C for 5 days in winter) (Hope and McElroy 1990). During the spring, when switchgrass has set shoots and opened buds it can resist temperatures of −5 °C for no more than 2 days (Hope and McElroy 1990). Long term field experiments in Bologna (Italy) showed a LGP $_{t5}$ amount from 190 to 220 days, accounting for a minimum of 2020 °C TS_5 in 2014 to a maximum of 2356 °C TS_5 in 2009 for Alamo (lowland ecotype), whereas two studies in Pisa coastal plain (Central Italy) assessed a TS_5 requirement of 2060 °C and 2540 °C for lowland and upland ecotype respectively (Lasorella 2014).

Switchgrass can better tolerate short growing seasons than giant reed, particularly considering that the upland ecotype is naturally occurring until 55°N latitude in America. Interestingly, in Europe switchgrass is supposed to grow even further north: Castillo et al. (2015) assessed that both switchgrass and giant reed can easily acclimate to the sub-severe LGP_{t5} threshold, which, according to the FAO Food Global Geographic Information System (GIS) Database (FFGGD) can reach 60°N latitude for switchgrass, and around 55°N for giant reed in Europe. Lowland switchgrass variety Kanlow yielded 17.54 Mg d.w. ha^{-1} in the third year after establishment in a sandy site in Wageningen (52° latitude), The Netherlands (Elbersen 1998).

5.2 Dryness

In order to assess dryness, the JRC experts chose the UNEP-AI (United Nations Environmental Program - Aridity Index) as a measurement system, i.e. the ratio between the annual precipitation and the potential evapotranspiration. The 'dry sub-humid' class, ranging between 0.5 and 0.65 (Middleton and Thomas 1997) was used to set the sub-threshold and threshold values, but was subsequently restricted to $0.5 \leq AI \leq 0.6$ by the Joint Research Commission (Jones et al. 2014). In Europe, the driest areas in AI are located in southern Europe, with south-east Spain (AI = 0.2–0.5) just within the limits of the 'semi-arid' zone. Spain is largely within the 'dry sub-humid' zone together with the southern part of Italy (including Sardinia), Greece, Poland and Hungary in Central Europe (Middleton and Thomas 1997).

Giant reed tolerance to drought is high (Christou et al. 2001; Rezk and Edany 1979). However, during establishment it is important to avoid drought, because the rhizomes require moist soils to sprout. The newly sprouted shoots can survive drought (Mann et al. 2013), and from the second year, drought tolerance can increase as the rhizome system can reach water sources in deep soil horizons. Productivity is, however, affected by drought as assessed in greenhouse studies carried out in California and Spain (Lambert et al. 2014; Sánchez et al. 2016). In addition, giant reed grown in dry conditions reached the yield ceiling later than in wet sites, but still maintained a good productivity (15.6 Mg d.w. ha^{-1} in a rainfed semi-arid environment in the Mediterranean), as shown in experiments in southern Italy (Alexopoulou et al. 2015) and confirmed by the rate of Water Use Efficiency (WUE) it can reach, ranging between 1.2 kg d.w. ha^{-1} mm^{-1} in the first year after the establishment to 7.6 kg d.w. ha^{-1} mm^{-1} in a mature plantation (Christou et al. 2003; Erickson et al. 2012; Mantineo et al. 2009; Triana et al. 2014).

Switchgrass is tolerant to drought; however, like for giant reed, the establishment in arid regions can be problematic due to frequently dry and/or coarse soils (Evers and Butler 2000; Evers and Parsons 2003) on which germination rates can be considerably reduced (Tucker et al. 2011). On the contrary, on clay soil, switchgrass showed very good survival rates in the first year after establishment, even under high water stress (Evers and Parsons 2003), suggesting that soil texture plays an important role in plant tolerance to drought at germination stage. The tolerance increases after a few years of establishment with survival rates of 100% for plants with an absorption pressure of −4 MPa (Barney et al. 2009). The upland ecotype has higher resistance to drought than lowland (Stroup et al. 2003; Porter 1966) and its WUE ranges from 1.6 to 5.6 kg d.w. ha^{-1} mm^{-1} (Zegada-Lizarazu et al. 2012).

5.3   Excess Soil Moisture

This constraint is calculated as number of days at or above field capacity, with a sub-severe threshold ranging from 210 to 230 days. In Europe, areas with excess soil moisture are mainly located in north-west Europe, where many soils exceed 300 days above field capacity (Confalonieri et al. 2014).

Being riparian plants, both giant reed and the lowland ecotype of switchgrass have high tolerance to excess soil moisture. They actually achieve higher yields when abundant water availability occurs in tandem with high temperatures (Lambert et al. 2014; Mann et al. 2013).

Soil trafficability and workability, as required by mechanization, could be hampered by excessively moist conditions on a clay soil, thus planting, harvesting or any other management activity (like mechanical weeding needed in the first year of establishment) can be hampered under these conditions. (Zegada-Lizarazu et al. 2010).

5.4   Limited Soil Drainage

The sub-severe threshold of this constraint are areas that are waterlogged for a significant duration of the year and are poorly drained.

Despite its adaptation to live along the shores of rivers and lakes and great tolerance to flooding and stagnating water, in a greenhouse experiment in California, under waterlogging conditions and in its first year after establishment, giant reed suffered a reduction of 47% and 40% in total and aboveground biomass production, respectively, compared to a field capacity moisture production (Lambert et al. 2014). Nevertheless, rhizomes' viability seems not to be affected by limited soil drainage (Mann et al. 2013).

The lowland ecotype of switchgrass has a higher tolerance to waterlogging than the upland ecotype. In a greenhouse experiment in California transplanted plants of the lowland ecotype performed as well in flooding conditions as in field capacity controls, whereas the upland ecotype reduced its biomass (Barney et al. 2009).

Although flooding is not a major constraint for the growth of these plants, agricultural mechanization can be impaired by poor drainage and can cause soil sealing. The sub-severe threshold limits for soil drainage are unlikely suitable for trafficability and workability as assessed by Castillo et al. (2015).

5.5 Unfavorable Texture and Stoniness

This is a broad biophysical constraint class that includes many soil groups with different relative abundances of clay, silt, sand, organic matter and coarse material.

Giant reed can grow in all types of soils from heavy clays to loose sands and gravelly soils (Perdue 1958), whereas switchgrass needs sand to clay loams and has low tolerance to heavy clay textures (Perdue 1958; Moser and Vogel 1995; Castillo et al. 2015).

No data concerning the growth of the two grasses in organic soils (with organic matter $\geq 30\%$) are available. Organic soils are mostly located in north-east Europe (Jones et al. 2005), a region that is popularly believed to be suitable for switchgrass cultivation. It is evident that further research is needed to assess the acclimation capacity of these species to soil texture.

5.6 Shallow Rooting Depth

The sub-severe threshold of 35–30 cm for shallow rooting depth is a severe constraint to the cultivation of both giant reed and switchgrass because of the progressive increase of the below ground rhizome network (Castillo et al. 2015). Shallow soil implies also limiting nutrients and water resources, as well as limited land workability. Giant reed can reach a remarkable height of 8–9 meters and produces a large aerial biomass (Angelini et al. 2009; Mantineo et al. 2009). A deep soil is needed to develop a proportional rhizome system, and to root stably in the substrate without lodging.

Switchgrass is shorter than giant reed and has thinner stems, but this sub-severe threshold is too shallow also for the rooting system of switchgrass (Castillo et al. 2015).

Shallow soils are most frequent in mountain and hill areas, which are frequent in Europe. Consequently shallow soils are often combined with slopes (Confalonieri et al. 2014).

5.7 Poor Chemical Properties (Salinity)

The sub-severe threshold for salinity is 3.2–4.0 dS/m (roughly 17–21 ppt). Saline soils are rapidly increasing in Europe due to climate warming, drought pressure, sea level raising and erroneous irrigation management.

Giant Reed is described as a halophyte plant with high resistance to salinity (Williams et al. 2009). In some studies carried out in South Australia, it produced 45.2 Mg d.w. ha^{-1} when irrigated with wastewater with a salinity of 9 dS/m (Williams et al. 2009).

In greenhouse experiments conducted in Illinois, upland switchgrass showed low tolerance to 5 dS/m (reduction of 77% in the aboveground dry weight), while the lowland ecotype proved higher tolerance to the same concentration (reduction of the 20% in the aboveground dry weight). However the salinity tolerance of 'Alamo' and 'Kanlow' (lowland ecotypes) was found to be intermediate with both having high germination percentages at 0 dS m^{-1}, but exhibiting reduction in germination percentages as salinity was increased to 20 dS m^{-1} (Anderson et al. 2015). In a growth chamber study the germination of 'Cave-in-Rock' (switchgrass upland ecotype) decreased by 15, 61 and 95% when subjected to NaCl solutions of 5, 10 and 15 dS m^{-1} respectively, when compared with the control (Carson and Morris 2012). Growth of 'Blackwell' switchgrass (upland ecotype) was not inhibited by high soluble salt levels present in anaerobically digested waste-activated sewage sludge, with the highest biomass yields being produced in pure sludge at EC 8.8 dS m^{-1} (Rodgers and Anderson 1995). Another study found that 'Blackwell' established from seed in acidic, saline bauxite mine soils, but provided poor groundcover and produced little biomass (Harper and Spooner 1983).

European saline soils are in the coastal areas of Greece, Italy, France, Portugal, north Germany and across Spain, which recorded the highest potentially saline areas (Tóth et al. 2008).

5.8 Poor Chemical Properties (Sodicity)

Sodicity is the percentage of the sodium cation adsorbed by the clay fraction of soils (Jones et al. 2014). Besides the effects on plant ecophysiology, exchangeable sodium excess affects also soil structure increasing erosion risk, vulnerability to waterlogging and availability of nutrients. The sub-severe threshold of 4.8–6 Exchangeable Sodium Percentage (ESP) in half or more than the 100 cm top layer is a severe constraint for the establishment of non-adapted perennial grasses.

Considering that the ESP is often determined using laborious and time consuming laboratory tests and that the Sodium Adsorption Ratio (SAR) is the most widely used index across the scientific literature it is useful to use a linear regression model developed by Rashidi and Seilsepour (2011) in order to equalize the data.

A high value of the SAR has a strong negative impact on soil water permeability and infiltration rates and can result in surface crusts that can negatively affect plant growth under field conditions (Brown et al. 1983). It has been observed that crops that are able to withstand excess moisture conditions, resulting in short-term oxygen deficiencies, are also more tolerant to sodic conditions because of the decreased water permeability of the soil (Abrol et al. 1988). Giant reed grown in wetlands characterized by a SAR of 11.8 (cmol $kg^{-1})^{0.5}$ (corresponding to around 14.1 ESP) did not exhibit any sodicity symptoms (Idris et al. 2012), and a greenhouse experiment in Illinois showed that switchgrass germination was not significantly affected by sodicity when the SAR was increased from 7 to 15 (cmol $kg^{-1})^{0.5}$ (corresponding to 9.2 and 17.4 ESP respectively) both for upland and lowland varieties (Anderson et al. 2015).

A macro analysis of European soils showed that south France and south Spain coastal areas are affected by sodicity, as well as large areas of Hungary, Romania and Bulgaria (Tóth et al. 2008).

5.9 Poor Chemical Properties (Topsoil Acidity, pH)

Topsoil acidity reduces nutrients availability in the soil solution and increases the concentration of some metals (i.e. aluminum) with toxic effects on the decomposing and nitrogen fixing communities (Spies and Harms 2007). The sub-severe threshold is between a pH of 5.5 and 5.

Giant reed performs best in the pH range between 5 and 7 (DiTomaso 1998). Other studies assessed that pH values from 4 to 5 are poorly suitable for giant reed, whereas the optimum is from 6 to 7 and a moderate suitability from 7 to 8 (Castillo et al. 2015).

Switchgrass has its optimal growth between pH 6 and 7, although soil acidity is not a limiting factor (Castillo et al. 2015; Wolf and Fiske 1995). Seedlings can tolerate acidic soils (Bona and Belesky 1992; Hopkins and Taliaferro 1997), and switchgrass roots have been reported to grow at pH 3.7 (Stucky et al. 1980). The major impact of acidic soil in plant growth is the associated phosphorus unavailability, as assessed in experiments carried out in Pennsylvania with switchgrass, which yielded 50% less biomass in strong acidic soils (pH 4.3–4.9) than in the control (Sanderson et al. 2012). Other studies assessed tolerance to topsoils within 4.9–7.6 pH values (Wolf and Fiske 1995), and Castillo et al. (2015) assessed a decreasing suitability for switchgrass from pH range 6–7 which scored the 'highest suitability' to pH range 5–6 'moderate suitability' and pH range 4–5 'poor suitability'.

Acidic topsoils are spread all over Europe, but the lowest pH values occur in central, eastern and northern Europe and in the Atlantic regions. Acidic topsoils are rare in the warm-temperate Mediterranean areas (Reuter et al. 2008).

5.10   Steep Slope

Steep slope is mainly a constraint for mechanization, and for the side effects of inclination like shallow soil, low water retention capacity, risk of landslide and soil depletion. Under particular conditions, slopes can result into positive effects: a south exposed slope can, for instance, have a better irradiance and higher temperatures than a plain, or north exposed, soil, however the soil moisture could be reduced. Slopes in the range 12–15% are within the sub-severe threshold.

Although giant reed (Impagliazzo et al. 2016) and switchgrass (Lee et al. 2009) growth is slightly affected by slope, the cultivation of these species could be prevented by the use of machines required for planting and harvesting. For instance, seed-drills are unsuitable for planting giant reed rhizomes and the development of specific equipment for the propagation of the plantlets combined with the high establishment cost could hamper the cultivation of giant reed in these areas. On the contrary a cereal like seed-drill can be used for switchgrass seeds (suitable even over 20% slope), and this could make switchgrass more attractive than giant reeds in slopy areas.

In Europe steep slopes are located in mountain and hill areas, hence quite uniformly spread across EU-28 (Confalonieri et al. 2014).

6   Suitability of Giant Reed and Switchgrass for "Unclear Synergies Output"

Of the pairwise combinations among biophysical constraints, "unclear synergies", as defined in the paragraph 3.2, are the ones to investigate, as 'negative synergies' are not suitable combinations for any agricultural activity, and 'positive synergies' drop any restriction to agriculture. The 'unclear synergies' can have positive, negative or unclear effects of one constraint on the other, and frequently depend on external unknown factors (i.e. north/south facing slopes), on the specific crops (i.e. different effects according to different species that are cultivated) or can act differently according to the specific situation (e.g. different outcome under a dry or wet climate). Hence the European regulation leaves the member states the possibility to demonstrate their own cases of ANC land for agriculture defined by 'unclear synergies' of biophysical constraints (Table 2) (Confalonieri et al. 2014).

Soil texture and stoniness are not severe constraints for rhizomatous plants like giant reed and switchgrass, which can establish in non-optimal soils and reach

Table 2 Unclear synergies of biophysical constraints for the determination of "areas with natural constraints" (ANC)

Biophysical constraints (1)	x	Biophysical constraints (2)	Giant reed	Switchgrass
Excess soil moisture	x	Unfavourable texture and stoniness (stoniness)	Very suitable	Suitable
Unfavourable texture and stoniness (organic soil)	x	Steep slope	Medium suitable	Suitable
Unfavourable texture and stoniness (stoniness)	x	Unfavourable texture and stoniness (heavy clay)	Suitable	Medium suitable
Unfavourable texture and stoniness (sand, loamy sand)	x	Unfavourable texture and stoniness (heavy clay)	Suitable	Medium suitable
Unfavourable texture and stoniness (heavy clay)	x	Steep slope	Medium suitable	Medium suitable

water and resources through their rhizome network. However, if organic soils and/or clay soils are combined with a steep slope, mechanization could be hampered by a reduced trafficability bearing strength and, as a consequence, the land may be unsuitable for agriculture. South exposed slopes can, on the other hand, have a positive effect on unfavourable soils and allow agriculture if the conditions are not extreme for mechanization. Organic soils on steep slopes occur in northern Europe in mountain and hilly areas (Jones et al. 2005). Heavy clay soils on steep slopes originate from in situ weathering of soils from basic parent material like marl, limestone, basalt. In Europe their occurrence is limited to mountain areas (Driessen et al. 2001). Perennial crops like giant reed and switchgrass could help protect these soils from erosion, however, since the establishment of switchgrass is easier and has a lower cost than that of giant reed, switchgrass could be the best technical option for slopy land. Stoniness, as such, has positive effects on clay and organic soils. It can, in fact, improve water percolation, soil aeration, soil warming and reduce the duration of wet periods. The improved water drainage, has on the other hand, an impact on the capacity of the soil to retain water and nutrients and thus on the availability of resources for the crops.

Excess moisture in stony soils results in conditions similar to those of rivers and lakes shores, i.e. the habitat of giant reed and the lowland ecotype of switchgrass, and are therefore not limiting the growth of these species. Hence the synergy of these two biophysical constrains turns into very favourable conditions for these plants. According to Confalonieri et al. (2014), these specific climate and soil characteristics are more frequent in northern and eastern Europe, where the cultivation of switchgrass could be preferable to that of giant reed for its adaptation to a short growing season.

Likewise, excess moisture, sandy or loamy sand soils occur with clay soils most often in the flood plains of rivers, and do not constitute limiting factors for giant reed and the lowland ecotype of switchgrass. A superficial layer of sand or loamy sand above a clay layer improves surface permeability, decreases waterlogging and increases the temperature of the topsoil, enhancing soil fertility and water availability. On the contrary, if the sand or loamy sand layer is deep in the ground and below a clay layer, fertility can be reduced. Although giant reed is more tolerant than switchgrass to heavy clay topsoils (Castillo et al. 2015; Perdue 1958), the order of the layers is not limiting the growth of both species as they can reach the layer with the most favourable conditions through rhizomes, in case that soil is deep enough for the development of a wide rhizome network.

7   Synthesis and Forward Outlook

Many areas across Europe are affected by one or more constraints limiting agriculture and, among these, drought is cause of extreme concern in the Mediterranean region, whereas low temperature is the main limiting factor in the regions of north and east Europe (Middleton and Thomas, 1997).

The target of the cultivation of perennial lignocellulosic crops is to produce high biomass, thus the low temperature, as defined by the JRC, does not necessarily limit the cultivation of perennial energy crops in ANC land with temperature constraints, in fact, even if the growing season is shorter than the optimal one, the only consequence could be a reduction in biomass yield since there is no need to complete the plant cycle itself and wait for fruits to mature to have a harvest. Essentially, killing frost is a more severe limiting factor than low temperature for these crops because it can lead to plant death and null yields. Studies assessed that for giant reed killing frost is -5 °C for 5 days in winter and 0 °C for 2 days in spring (El Bassam 2010), whereas for switchgrass it is -20 °C for 5 days in winter and -5 °C for 2 days in spring (Hope and McElroy 1990).

After killing frost, the most serious constraints for the cultivation of giant reed and switchgrass in ANC land appear to be a shallow soil and limited soil drainage (Table 1). Shallow soils do not allow plants to root stably and develop a rhizome network suitable to find resources to flourish in ANC land. Land that is waterlogged for a long time (sub-severe threshold) is not limiting the growth of these two species since both, giant reed and the lowland ecotype of switchgrass, are riparian plants and can benefit from waterlogging but this can hamper mechanization and the possibility of cultivating wetlands in conventional ways. Giant reed and switchgrass wide ecological amplitude and the development of a wide subterranean rhizome network suggest that organic soils are not a constraint for giant reed and switchgrass. However, lack of data on the growth of the two species in organic soil leaves an open conundrum about the acclimation capacity of these plants to these soils which needs to be solved with targeted studies.

With the exception of the three above mentioned limiting factors, both switchgrass and giant reed can potentially establish and grow well in ANC land, with no or low input. However, these conclusions are based on a literature of growth experiments mostly conducted with pot plants and in greenhouses. Data recorded in open field experiments in land with exactly such specific constraints or combinations of such constraints (as indicated in the Rural Development Regulation) are scarce, but are focal to assess the effective response of these plants to these conditions and evaluate if the resulting yields can guarantee an income for the farmer and stimulate the development of a supply chain and of a market. Economically speaking, it is likely that in some ANC lands, switchgrass and giant reed productivity is too low to justify a rationale land allocation. Low productivity means an insufficient feedstock for the transformation industry that, instead, needs a continuous inflow of lignocellulosic biomass. Likewise, the absence of infrastructures (typical in ANC land) and the distance to the transformation plant are relevant hurdles that affect the diffusion of a sustainable large-scale value chain. Martelli et al. (2015) and Mitchell et al. (2012) calculated that transformation plants should be within a distance of approximately 40 km from the fields to ensure sustainability.

Another aspect which needs to be assessed along with the technical feasibility of supplying energy feedstock produced in ANC land and the sustainability of the value chain, is the environmental impact and sustainability of such crops. The cultivation of ANC land with perennial lignocellulosic crops can provide important ecosystem services which should be considered and pursued along with the production of biofuel feedstock. Such services are protection of soil from erosion (by increasing vegetation cover), improvement of soil conditions and fertility (SOM and N increase due to the absence of soil tillage) and climate change mitigation (C storage in the soil) (Fagnano et al. 2015). The use of ANC land and the introduction of lignocellulosic crops into new ranges has, however, also drawbacks which need to be understood to fully assess the sustainability of cultivating ANC land with bioenergy crops. ANC land is a valuable source of biodiversity and has a high nature conservation potential, both in anthropogenic, agricultural and wild environments. The establishment of monospecific crops affects biotic communities and nutrient cycles, and alters ecosystems dynamics and the natural succession of biodiversity, even if the cultivation is only temporary. In addition, plants like the giant reed and switchgrass, that grow fast and produce large yields, have high tolerance to stress and low input requirements, are very competitive and can easily escape cultivation and naturalize in new ranges, and eventually become invasive (Barney 2014). Both giant reed and switchgrass are not native to Europe and giant reed is shortlisted among the ten most invasive species in the world (Lowe et al. 2000). The cultivation of these species as crops contributes to the dispersal of their propagules, and can boost invasive spreading, feeding back with landscape and ecosystems loss, GHG emissions, management costs and loss of societal opportunities (Marabuah et al. 2014). Therefore research needs to address also the evolution of ANC land and the effects of the cultivation of species like giant reed and switchgrass in order to define criteria for the allocation of ANC land to agriculture rather than nature, regulate the introduction and establishment of carbon neutral biofuel crops and define agronomic measure to control their dispersal.

The sustainability of the cultivation of perennial lignocellulosic crops on ANC land, should represent the keystone requirement for the eligibility of CAP funding

Table 3 Projection of released land from agriculture in 2020 and 2030 in Europe, potentially available for lignocellulisc biomass production

Reference	2020 (Mha)	2030 (Mha)	Criterion
Krasuska et al. (2010)	20.5	26.3	UAA ceasing to be suitable for food/ feed production.
EEA (2007)		20	Land released from agricultural production, set aside lands used for bioenergy purpose (environmental constraint).
Fischer et al. (2010)		22–46	('LU-env' and 'LU-ene' respectively) released land from intensive agriculture.
Ramirez-Almeyda et al. (2017)	17.4–31.3	18.7–32.4	Released land from UAA (including orchards, olive groves, vineyards and fallow land) excluding high nature value land (HNV), permanent grassland and Nature2000 sites.
Valin et al. (2015)		13	Decrease in traditional cropland to the benefit of energy plantations for solid biomass, afforestation and other abandoned areas.

and the development of supply chains, and should be assessed considering also the environmental impact of ANC land use and of the cultivation of highly producing species, and not only their GHG emissions.

The projections of land from ceased agriculture predict that from 17 to 46 Mha will potentially be available for bioenergy feedstock production between 2020 and 2030 in Europe (Table 3). These estimates depend on the criteria used to define available land for biofuel crops and the European countries considered.

Based on these projections, large areas of potentially available land for the production of biofuel feedstock will be released from agriculture in the next few years. The S2BIOM project (Ramirez-Almeyda 2017; Ramirez-Almeyda et al. 2017) showed that most of such land will be released in southern Europe (Spain and Turkey) and east Europe (Ukraine and Romania), in areas where giant reed and switchgrass would thrive well, as seen in the present study.

However, given the several uncertainties associated with the productivity and profitability of ANC land, the development of a value-chain in these areas and the environmental issues associated with the use of ANC land for biofuel crops, it is wise to consider alternative biomass sources both in terms of land and species, which can integrate and release the pressure on ANC productivity. It is with this intent that the European BECOOL project (Brazil-EU Cooperation for Development of Advanced Lignocellulosic Biofuels) is testing innovative cropping systems and soil ameliorating species (like leguminous plants) by integrating food with biomass crops through innovative crop rotation systems, and increasing the Land Use Ratio (LUR) in high productive lands.

Less generalist and invasive species than giant reed and switchgrass, might also be a productive option to be evaluated under the specific limiting conditions of ANC land. A projection of ANC land in Europe as defined by the above mentioned biophysical constraints is missing. It would be particularly interesting to overlap

ANC projections with those of land use, released land from agriculture and nature protection in order to identify more precisely the areas, the potential species and the economic and environmental issues to be addressed with future research.

References

Abrol IP, Yadav JSP, Massoud FI (1988) Salt-affected soils and their management. Food and Agriculture Organization of the United Nations

Alexopoulou E, Zanetti F, Scordia D, Zegada-Lizarazu W, Christou M, Testa G, Cosentino SL, Monti A (2015) Long-term yields of switchgrass, giant reed, and miscanthus in the mediterranean basin. Bioenergy Res 8(4):1492–1499. https://doi.org/10.1007/s12155-015-9687-x

Anderson EK, Voigt TB, Kim S, Lee DK (2015) Determining effects of sodicity and salinity on switchgrass and prairie cordgrass germination and plant growth. Ind Crop Prod 64:79–87. https://doi.org/10.1016/j.indcrop.2014.11.016

Angelini LG, Ceccarini L, Nassi O, Di Nasso N, Bonari E (2009) Comparison of *Arundo donax* L. and *Miscanthus x giganteus* in a long-term field experiment in Central Italy: analysis of productive characteristics and energy balance. Biomass Bioenergy. https://doi.org/10.1016/j.biombioe.2008.10.005 33:635

Bacher W, Sauerbeck G (2001) Giant Reed (*Arundo donax* L.) Network Improvement biomass quality Final report FAIR-CT-96-2028. Braunschweig Bundesforschungsanstalt Für Landwirtschaft (FAL)

Barney JN (2014) Bioenergy and invasive plants: quantifying and mitigating future risks. Invasive Plant Sci Manag 7:199–209

Barney JN, Mann JJ, Kyser GB, Blumwald E, Van Deynze A, Ditomaso JM (2009) Tolerance of switchgrass to extreme soil moisture stress: ecological implications. Plant Sci. https://doi.org/10.1016/j.plantsci.2009.09.003 177:724

El Bassam N (2010) Handbook of bioenergy crops: a complete Refrence to species, development and applications, earthscan. London P 545:140

Bona L, Belesky DP (1992) Evaluation of switchgrass entries for acid soil tolerance. Commun Soil Sci Plant Anal 23:1827–1841

Bonet A (2004) Secondary succession of semi-arid Mediterranean old-fields in South-Eastern Spain: insights for conservation and restoration of degraded lands. J Arid Environ 56(2):213–233. https://doi.org/10.1016/S0140-1963(03)00048-X

Brown PL, Halvorson AD, Siddoway FH, Mayland HF, Miller R (1983) Saline-seep diagnosis, control, and reclamation. In: Report # 30. United States Department of Agriculture, Agricultural Research Service

Carson MA, Morris AN (2012) Germination of *Panicum virgatum* cultivars in a NaCl gradient. Bios 83:90–96

Castillo CP, Lavalle C, Baranzelli C, Mubareka S (2015) International journal of geographical information science modelling the spatial allocation of second-generation feedstock (lignocellulosic crops) in Europe Modelling the spatial allocation of second-generation feedstock (lignocellulosic crops) in Europe. Int J Geogr Inf Sci. https://doi.org/10.1080/13658816.2015.1051486 29:1807

Christou M, Mardikis M, Alexopoulou E (2001) Research on the effect of irrigation and nitrogen upon growth and yields of *Arundo donax* L. in Greece. Aspects Appl Biol 65, Biomass and energy crops:47–55

Christou M, Mardikis M, Alexopoulou E, Cosentino SL, Copani V, Sanzone E (2003) Environmental studies on *Arundo donax*. 8th International Conference on Environmental Science and Technology Lemnos Island, Greece, 8–10 September 2003, (September), 102–110. Available at: http://www.cres.gr/bioenergy_chains/files/pdf/Articles/3-Lemnos.pdf. Accessed 05 May 2017

Confalonieri R, Jones B, Van Diepen K, Van Orshoven J (2014) Scientific contribution on combining biophysical criteria underpinning the delineation of agricultural areas affected by specific constraints methodology and factsheets for plausible criteria combinations

DiTomaso JM (1998) Biology and ecology of giant reed. 1–5. In: Bell C (ed) Arundo and saltcedar: the deadly duo. Proceedings of the Arundo and saltcedar workshop University of California Cooperative Extension Publication. Imperial County, CA

Driessen P, Deckers J, Spaargaren O, Nachtergaele F (eds) (2001) Lecture notes on the major soils of the world. FAO world soil resources reports no. 94. 334 pp. FAO, Rome

Elbersen IHW (1998) Switchgrass (*Panicum virgatum* L.) as an alternative energy crop in Europe Initiation of a productivity network. Final Report for the period. FAIR 5-CT97-3701 http://www.switchgrass.nl/upload_mm/6/3/7/c9842903-b468-436a-8412-9f47c296b2de_Sw_FinalRep_full2.pdf

Erickson JE, Soikaew A, Sollenberger LE, Bennett JM (2012) Water use and water-use efficiency of three perennial bioenergy grass crops in Florida. Agriculture 2(4):325–338. https://doi.org/10.3390/agriculture2040325

European Environmental Agency (EEA) (2007) Estimating the environmentally compatible bioenergy potential from agriculture. European Environmental Agency https://doi.org/10.2800/13734

European Environment Agency (2008) Energy and environment report 2008. Energy (Vol. 6/2008). https://doi.org/10.2800/10548

Evers GW, Butler TW (2000) Switchgrass establishment on coastal plain soil. In: Proc. Amer.Forage Grassl. Council. pp. 150–154. Madison, WI, July 16–19, 2000. Georgetown, TX

Evers GW, Parsons MJ (2003) Soil type and moisture level influence on Alamo switchgrass emergence and seedling growth. Crop Sci 43:288–294

Fagnano M, Impagliazzo A, Mori M, Fiorentino N (2015) Agronomic and environmental impacts of giant reed (*Arundo donax* L.): results from a long-term field experiment in hilly areas subject to soil Erosion. Bioenergy Res 8(1):415–422. https://doi.org/10.1007/s12155-014-9532-7

FAOSTAT (2012) Food and agriculture organization of the United Nations. Available at: http://faostat3.fao.org/. Accessed 15 April 2017

Fischer G, Prieler S, van Velthuizen H, Berndes G, Faaij A, Londo M, de Wit M (2010) Biofuel production potentials in Europe: sustainable use of cultivated land and pastures, part II: land use scenarios. Biomass Bioenergy 34(2):173–187. https://doi.org/10.1016/j.biombioe.2009.07.009

Food and Agriculture Organization & International Institute for Applied Systems Analysis (2000) Coefficient of variation of length of growing period, 1901–1996. Global agro-ecological zones. In FAO & IIASA, 2007., Mapping biophysical factors that influence agricultural production and rural vulnerability

Hardion L, Verlaque R, Saltonstall K, Leriche A, Vila B (2014) Origin of the invasive *Arundo donax* (*Poaceae*): a trans-Asian expedition in herbaria. Ann Bot 114(3):455–462. https://doi.org/10.1093/aob/mcu143

Harper J, Spooner AE (1983) Establishment of selected herbaceous species on acid bauxite minesoils. In: Proc. 1983 Symposium on Surface Mining, Hydrology, Sedimentology and Reclamation, pp. 413–422

Hope HJ, Mcelroy A (1990) Low-temperature tolerance of switchgrass (*Panicum virgatum* L.). J Plant Sci Can J Plant Sci. Downloaded from 70(109):109–1096. Available at: www.nrcresearchpress.com. Accessed 25 April 2017

Hopkins AA, Taliaferro CM (1997) Genetic variation within switchgrass populations for acid soil tolerance. Crop Sci 37:1719–1722

Hsu FH, Nelson CJ, Matches AG (1985) Temperature effects on germination of perennial warm-season forage grasses. Crop Sci 25:215–220

Idris SM, Jones PL, Salzman SA, Croatto G, Allinson G (2012) Evaluation of the giant reed (*Arundo donax*) in horizontal subsurface flow wetlands for the treatment of recirculating aquaculture system effluent. Environ Sci Pollut Res 19(4):1159–1170. https://doi.org/10.1007/s11356-011-0642-x

Impagliazzo A, Mori M, Fiorentino N, Di Mola I, Ottaiano L, De Gianni D, Fagnano M (2016) Crop growth analysis and yield of a lignocellulosic biomass crop (*Arundo donax* L.) in three marginal areas of Campania region. Ital J Agron 11:1–7. https://doi.org/10.4081/ija.2016.755

Jones RJA, Hiederer R, Rusco E, Loveland PJ, Montanarella L (2005) Estimating organic carbon in the soils of Europe for policy support. European Journal of Soil Science 2005(56):655–671

Jones R, Le-Bas C, Nachtergaele F, Rossiter D, Van Orshoven J, Schulte R, Van Velthuizen H (2014) Updated common bio-physical criteria to define natural constraints for agriculture in Europe. Definition and scientific justification for the common criteria. JRC Science and Policy Reports. EUR 26638 EN, 68. https://doi.org/10.2788/79958

Krasuska E, Cadórniga C, Tenorio JL, Testa G, Scordia D (2010) Potential land availability for energy crops production in Europe. Biofuels Bioprod Biorefin. https://doi.org/10.1002/bbb.259 4:658

Lambert AM, Dudley TL, Robbins J (2014) Nutrient enrichment and soil conditions drive productivity in the large-statured invasive grass *Arundo donax*. Aquat Bot 112:16–22. https://doi.org/10.1016/j.aquabot.2013.07.004

Lasorella MV (2014) Suitability of Switchgrass (*Panicum virgatum* L.) Cultivars in Mediterranean Agroecosystems

Lee DK, Owens VN, Boe A, Koo BC (2009) Biomass and seed yields of big bluestem, switchgrass, and intermediate wheatgrass in response to manure and harvest timing at two topographic positions. GCB Bioenergy 1:171–179. https://doi.org/10.1111/j.1757-1707.2009.01008.x

Lewandowski I, Scurlock JMO, Lindvall E, Christou M (2003) The development and current status of perennial rhizomatous grasses as energy crops in the US and Europe. Biomass Bioenergy 25:335–361. https://doi.org/10.1016/S0961-9534

Lowe S, Browne M, Boudjelas S, De Poorter M (2000) 100 of the world's worst invasive alien species. A selection from the global invasive species database. The Invasive Species Specialist Group (ISSG) of the Species Survival Commission (SSC) of the World Conservation Union (IUCN)

Mann JJ, Kyser GB, Barney JN, DiTomaso JM (2013) Assessment of aboveground and below-ground vegetative fragments as Propagules in the bioenergy crops *Arundo donax* and *Miscanthus × giganteus*. Bioenergy Res 6(2):688–698. https://doi.org/10.1007/s12155-012-9286-z

Mantineo M, D'Agosta GM, Copani V, Patanè C, Cosentino SL (2009) Biomass yield and energy balance of three perennial crops for energy use in the semi-arid Mediterranean environment. Field Crop Res. https://doi.org/10.1016/j.fcr.2009.07.020 114:204

Marabuah G, Gren I, McKie B (2014) Economics of harmful invasive species: a review. Diversity 6:500–523. https://doi.org/10.3393/d6030500

Mariani C, Cabrini R, Danin A, Piffanelli P, Fricano A, Gomarasca S, Soave C (2010) Origin, diffusion and reproduction of the giant reed (*Arundo donax* L.): a promising weedy energy crop. Ann Appl Biol 157(2):191–202. https://doi.org/10.1111/j.1744-7348.2010.00419.x

Martelli R, Bentini M, Monti A (2015) Harvest storage and handling of round and square bales of giant reed and switchgrass: an economic and technical evaluation. Biomass Bioenergy 83:551–558. https://doi.org/10.1016/j.biombioe.2015.11.008

Middleton NJ, Thomas D (1997) World atlas of desertification. UNEP, Arnold, London

Mitchell R, Vogel KP, Uden DR (2012) The feasibility of switchgrass for biofuel production. Biofuels 3:47–59. https://doi.org/10.4155/bfs.11.153

Monti A, Zatta A (2009) Root distribution and soil moisture retrieval in perennial and annual energy crops in Northern Italy. Agric Ecosyst Environ 132(3):252–259. https://doi.org/10.1016/j.agee.2009.04.007

Monti A, Zegada-Lizarazu W (2016) Sixteen-year biomass yield and soil carbon storage of Giant reed (*Arundo donax* L.) grown under variable nitrogen fertilization rates. Bioenergy Res. https://doi.org/10.1007/s12155-015-9685-z 9:248

Moser LE, Vogel KP (1995) Switchgrass, big bluestem, and indiangrass. In: Barnes RF, Miller DA, Nelson CJ (eds) An introduction to grassland agriculture, Forages, vol 1, 5th edn. Iowa State Univ. Press, Ames, pp 409–420

Nassi o Di Nasso N, Roncucci N, Triana F, Tozzini C, Bonari E (2011) Productivity of giant reed (Arundo donax L.) and miscanthus (Miscanthus x giganteus Greef et Deuter) as energy crops: growth analysis. Ital J Agron 6:e22. https://doi.org/10.4081/ija.2011.e22

Nocentini A, Di Virgilio N, Monti A (2015) Model simulation of cumulative carbon sequestration by switchgrass (*Panicum virgatum* L.) in the mediterranean area using the DAYCENT model. Bioenergy Res. https://doi.org/10.1007/s12155-015-9672-4 8:1512

Perdue RE (1958) Arundo donax-source of musical reeds and industrial cellulose. Econ Bot 12(4):368–404. https://doi.org/10.1007/BF02860024

Porter CL Jr (1966) An analysis of variation between upland and lowland switchgrass *Panicum virgatum* L. in Central Oklahoma. Ecology 47:980–992

Ramirez Almeyda J (2017) Lignocellulosic crops in Europe: Integrating crop yield potentials with land potentials. PhD Thesis, University of Bologna, Italy, Department of Agricultural Science

Ramirez-Almeyda J, Elbersen B, Monti A, Staritsky I, Panoutsou K, Alexopoulou E, Schrijver R, Elbersen W (2017) Assessing the potentials for non food crops. In: Panoutsou K (ed) Modeling and optimization of biomass supply chains. Top-down and Bottom-up assessment for agricultural, forest and waste feedstock. Academic Press, Elsevier, London

Rashidi M, Seilsepour M (2011) Prediction of soil sodium adsorption ratio based on soil electrical conductivity. Middle East Journal of Scientific Research 8(2):379–383. Retrieved from http://www.idosi.org/mejsr/mejsr8(2)11/10.pdf

Reuter HI, Rodriguez Lado L, Hengl T, Montanarella L (2008) Continental-scale digital soil mapping using european soil profile data: soil ph. Available at: http://esdac.jrc.ec.europa.eu/public_path/shared_folder/dataset/10_soil_ph/soil_ph_in_europe_hbpl19_10.pdf. Accessed 06 April 2017

Rezk MR, Edany TY (1979) Comparative responses of two reed species to water table levels. Egypt J Bot 22(2):157–172

Rodgers CS, Anderson RC (1995) Plant-growth inhibition by soluble salts in sewage sludge-amended mine spoils. J Environ Qual 24:627–630

Sánchez J, Curt MD, Fernández J (2016) Approach to the potential production of giant reed in surplus saline lands of Spain. GCB Bioenergy, 1–14. https://doi.org/10.1111/gcbb.12329

Sanderson MA, Reed RL, McLaughlin SB, Wullschleger SD, Conger BV, Parrish DJ, Wolf DD, Taliaferro CM, Hopkins AA, Ocumpaugh WR, Hussey MA, Read JC, Tischler CR (1996) Switchgrass as a sustainable bioenergy crop. Bioresour Technol 56:83–93

Sanderson M A, Schmer M, Owens V, Keyser P, Elbersen W (2012) Crop management of switchgrass. Green Energy and Technology. https://doi.org/10.1007/978-1-4471-2903-5_4

Sharma KP, Kushwaha SPS, Gopal B (1998) A comparative study of stand structure and standing crops of two wetland species, *Arundo donax* and *Phragmites karka*, and primary production in *Arundo donax* with observations on the effect of clipping. Tropical Ecology 39(1):39

Spies CD, Harms CL (2007) Soil acidity and liming of Indiana soils AY-267 RR 6/88. Department of Agronomy, Purdue University. Purdue University. Cooperative Extension Service. West Lafayette

Stroup JA, Sanderson MA, Muir JP, Mc Farland MJ, Reed RL (2003) Comparison of growth and performance in upland and lowland switchgrass types to water and nitrogen stress. Bioresour Technol 86:65–72

Stucky DJ, Bauer JH, Lindsey TC (1980) Restoration of acidic mine spoils with sewage sludge: I. Revegetation. Reclamation Review 3:129–139

Terres JM, Nisini L, Anguiano E (2013) Assessing the risk of farmland abandonment in the EU Final report. https://doi.org/LB-NA-25783-EN-N

Tóth G, Montanarella L, Rusco E (2008) Threats to Soil Quality in Europe. Available at: http://eusoils.jrc.ec.europa.eu/ESDB_Archive/eusoils_docs/other/EUR23438.pdf. Accessed 05 May 2017

Triana F, Nassi O, Di Nasso N, Ragaglini G, Roncucci N, Bonari E (2014) Evapotranspiration, crop coefficient and water use efficiency of giant reed (*Arundo donax* L.) and miscanthus (*Miscanthus giganteus* Greef et Deu.) in a Mediterranean environment

Tucker SS, Craine JM, Nippert JB (2011) Physiological drought tolerance and the structuring of tallgrass prairie assemblages. GCB Bioenergy 7:811–819. Ecosphere, 2(4), art48. https://doi.org/10.1890/ES11-00023.1

Valin H, Peters D, van den Berg M, Frank S, Havlik P, Forsell N, Hamelinck C (2015) The land use change impact of biofuels in the EU: Quantification of area and greenhouse gas impacts, pp. 261. Available at: https://ec.europa.eu/energy/sites/ener/files/documents/Final Report_GLOBIOM_publication.pdf. Accessed 06 April 2017

Weaver JE (1968) Prairie plants and their environment: a fifty-year study in the Midwest. University of Nebraska Press, Lincoln, NE, p 276

Williams CMJ, Biswas TK, Black ID, Marton L, Czako M, Harris PL, Virtue JG (2009) Use of poor quality water to produce high biomass yields of giant reed (*Arundo donax* L.) on marginal lands for biofuel or pulp/paper. Acta Horticulturae 806:595–602

Wolf DD, Fiske DA (1995) Planting and managing switchgrass for forage, wildlife, and conservation. Virginia Cooperative Extension Pub. No. 418e013, Blacksburg, VA, USA

Wullschleger SD, Davis EB, Borsuk ME et al (2010) Biomass production in switchgrass across the United States: database description and determinants of yield. Agron J 102:1158–1168. https://doi.org/10.2134/agronj2010.0087

Zegada-Lizarazu W, Monti A (2011) Energy crops in rotation. A review. Biomass Bioenergy 35:12–25. https://doi.org/10.1016/j.biombioe.2010.08.001

Zegada-Lizarazu W, Elbersen HW, Cosentino SL, Zatta A, Alexopoulou E, Monti A (2010) Agronomic aspects of future energy crops in Europe. Biofuels Bioprod Biorefin. https://doi.org/10.1002/bbb.242 4:674

Zegada-Lizarazu W, Wullschleger S, Surendran Nair S, Monti A (2012) Crop physiology. In: Monti A (ed.) Switchgrass. A valuable biomass crop for energy. Springer, London, pp. 55–86

Zegada-Lizarazu W, Parrish D, Berti M, Monti A (2013) Dedicated crops for advanced biofuels: consistent and diverging agronomic points of view between the USA and the EU-27. Biofuels Bioprod Biorefin 7:715–731

The Availability and Economic Analyses of Using Marginal Land for Bioenergy Production in China

Yuqi Chen, Xiubin Li, Xudong Guo, and Chunyan Lv

Abstract Using marginal land for bioenergy production was considered an important strategy to promote this clean energy in China. In order to analyze the availability and economic feasibility of this strategy, this study used the latest official land use data to assess the bioenergy potentials of five species of energy plants on the marginal land, including *Manihot esculenta, Jatropha curcas, Helianthus tuberous L, Pistacia chinensis*, and *Xanthoceras sorbifolia Bunge*. The results indicate that 289.71 million ha of marginal land are available to cultivate these five energy plants and can produce 24.45 million tons bioethanol and 8.77 million tons of biodiesel. Based on field survey data and literature reviews, we found that both farmers and bioenergy plants have less enthusiasm for bioenergy production due to poor economic returns, albeit sufficient marginal land for energy plants.

Keywords Bioenergy · Marginal land · Energy plants · Economic benefits · China

1 Introduction

In recent years, China has witnessed rapidly increasing dependence on foreign oil imports (Chang et al. 2012). In 2015, the primary energy consumption in China was 543 million tons, of which 328 million tons was imported. The share of imported foreign oil increased from 49.8% in 2008 to 60.41% in 2015 (Fig. 1). To address the national energy security and Greenhouse Gas (GHG) emission reduction, China has made considerable efforts in expanding renewable energy portfolio, especially liquid biofuels (Zhao et al. 2015). For example, the yields of bioethanol and

Y. Chen (✉) · X. Li · X. Guo · C. Lv
China Land Surveying and Planning Institute, Key Laboratory of Land Use, Ministry of Land and Resource, Beijing, China

© Springer International Publishing AG, part of Springer Nature 2018
R. Li, A. Monti (eds.), *Land Allocation for Biomass Crops*,
https://doi.org/10.1007/978-3-319-74536-7_4

65

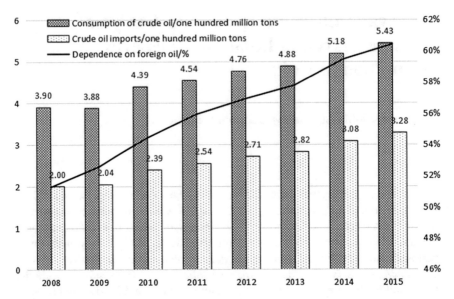

Fig. 1 Statistics of crude oil consumption, imports and dependence on foreign oil

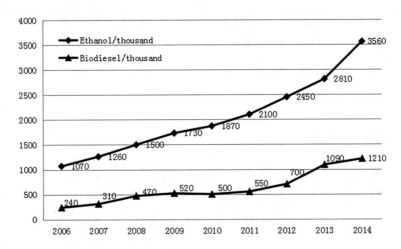

Fig. 2 Production of ethanol and biodiesel in China

biodiesel dramatically increased from 1.07 million and 0.24 million tons in 2006 to 3.56 million and 1.21 million tons in 2014 respectively (Fig. 2) (Qiu and Huang 2008).

However, under the pressure of high population and vulnerable food security, China's National Development and Reform Commission (NDRC) ruled that bioenergy production should focus on non-cereal feedstock (Li 2008). Energy plants can only be cultivated on marginal land, which is unsuitable for growing field crops due to edaphic and/or climatic limitations, vulnerability to erosion, or other environmental risks, but might be usable for growing energy plants (Wang and Shi 2015; Gelfand et al. 2013; Kang et al. 2013). This policy can potentially help avoid the side effects of expanding biofuel outputs from primary cropland (Swinton et al. 2011).

There have been a number of studies on using the marginal land for energy plants' cultivation in China. For example, by using of the field survey data of their own, Tang et al. (2010) proposed that if all the waste land, land boundary, road side land, stream side land, house surroundings, land along highways/roads can be regarded as marginal land, the area of marginal land was about 34.7 million ha. Based on the land use data from the Data Center for Resources and Environmental Sciences, Zhuang et al. (2011) used multi-factor analysis method and concluded that total area of marginal land suitable for energy plants was 43.75 million ha. Fu and Wu (2014) used the data from the Ministry of Agriculture of China and the Ministry of Land and Resources, and concluded the marginal land suitable for energy crops planting was 6.61 million ha. Because of the different definition of marginal land, different sources of land use data and different energy plants as research subjects, the results of total marginal land area varied dramatically. In order to figure out the accurate amount of marginal land, a reliable and reasonable result is necessary.

In addition, there has been plenty of marginal land suitable for energy plants' cultivation, economic viability of cultivating energy plants on marginal land is still largely unknown. There were very few relevant studies conducted to address both land availability and economic viability. The critical question to be addressed by this article is whether the current bioenergy strategy is feasible. This paper begins with a nation-wide statistical analysis of marginal land using the latest official land use data, presents a spatial analysis of this land availability, and then analyzes the willingness and economic benefits of farmers to cultivate the energy crops on the marginal land.

2 Natural Suitability of Marginal Lands for Biofuels Production

To estimate the potentials of biofuels production on marginal lands, the quantity and distribution of marginal lands were first determined. However, partially due to the broad country territory and high complexity in land use structure, there are only two official national land survey datasets, which were dated in 1997 and 2009 respectively. As a result, almost all of the potential energy production assessments on marginal land were based on relatively coarse land use data released by the Ministry of Land Resources, or human-interpreted land use maps. The difference in data sources could potentially result in inconsistent output data quality.

2.1 Status of Marginal Land in China

The latest and most authoritative land use data in China was from the 2nd National Land Survey. This national survey was organized by the Ministry of Land and Resources, and officially launched on the July 1st, 2007 and finished on December 31st, 2009. More than 200,000 leaders and staff participated in this work. Compared with previous effort, this land survey employed an unified national standard land use classification along with nationwide remote sensing based maps, including aerial photography images, the QuickBird satellite images, the Spot 5 panchromatic satellite images, and some Chinese local resources satellite images. The scale of rural land survey and urban land survey were mainly 1: 10000 and 1: 500 respectively. It can accurately reflect the true status of land use, and all data was finally published in the end of 2009. In addition, during the project implementation process, there were strict precision requirements during the Document Object Model production and image interpretation, and combined with field verification, which ensured the accuracy of the Survey data.

According to the 2nd National Land Survey, land can be generally classified into the Level I categories, including arable land, garden land, forestry land, grass land, residential and independent industrial & mining land, transportation land, waters and water conservancy facility land, and other land (such as free land, facility agricultural land, raised path through fields, saline land, wetland, sandy land, bare land). The area of these types of lands were 135.38 million ha, 14.81 million ha, 253.95 million ha, 287.31 million ha, 28.74 million ha, 7.94 million ha, 42.69 million ha and 176.98 million ha respectively in 2009 (Fig. 3)[1].

Under Level I, there are a total of 57 Level II classes. Considering the national strategies on arable land protection and ecological civilization, it is commonly agreed that the land with poor natural conditions but certain production potential and development values can be regarded as marginal land (Zhang et al. 2012; Zhuang et al. 2011). These lands include shrub land, bare land, inland beach, other grassland, other woodland, sandy land, coastal beach, saline land, wetland, totally 289.71 million ha. Figure 4 shows the spatial distribution of these lands in the nation's mainland territory.

2.2 Suitable Energy Plants for Biofuels in China

In China, in spite of thousands of species of oil plants, the species that can be cultivated on the barren land on a large scale are quite limited. Following the national policies that food, cooking oil and sugar are not allowed for producing fuels and farm lands are prohibited from cultivating energy plants, we selected five species of energy plants, *Manihot esculenta, Jatropha curcas, Helianthus tuberous L, Pistacia chinensis, Xanthoceras sorbifolia Bunge* based on the literature and field survey data (Chen et al. 2016; Zhuang et al. 2011).

---

[1]Since the base map in this study is the 1:100 million downsizing data, there may be a slight difference from the actual area.

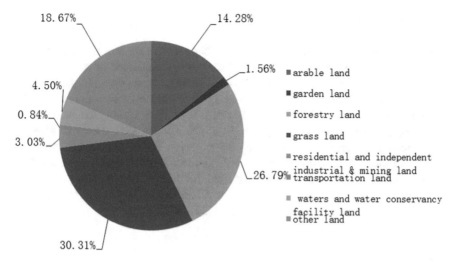

18. 67%   14. 28%

                    1. 56%   ■ arable land
4. 50%   ■ garden land
                             ▪ forestry land
0. 84%   ▪ grass land
3. 03%   ▪ residential and independent
                    26. 79% industrial & mining land
                             transportation land
                             ▪ waters and water conservancy
                               facility land
                             ▪ other land

30. 31%

Fig. 3 Land use status in 2009 in China

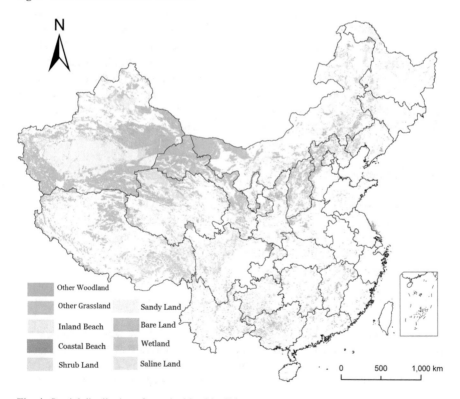

N

Other Woodland

Other Grassland   Sandy Land

Inland Beach   Bare Land

Coastal Beach   Wetland

Shrub Land   Saline Land

0   500   1,000 km

Fig. 4 Spatial distribution of marginal land in China

Manihot esculenta

Manihot esculenta is a woody shrub of the *Euphorbiaceae*. It is drought-tolerant, disease-resistant, easy to cultivation, and extensively cultivated as an annual crop in tropical and subtropical regions, including Guangxi, Guangdong, Hainan, Fujian, Yunnan and 13 other provinces. Fresh *Manihot esculenta* fruits contains 32–35% starch which can be used for producing ethanol. In 2006, China National Cereals, Oils and Foodstuffs Corp. (COFCO) constructed a 200,000 tons/year *Manihot esculenta*-based fuel ethanol demonstration project in Beihai, Guangxi Province. In 2014, the yield of *Manihot esculenta* in China was 4.66 million tons (FAOSTAT 2014). *Manihot esculenta* was mainly used for human consumption, livestock, industrial alcohol, poultry feed and chemical industry. After the establishment of COFCO, the contradiction for raw materials intensified.

Jatropha curcas

Jatropha curcas is a species of semi-evergreen tree in the spurge family and native to the tropical America. It is now widely cultivated in Southwest China, such as Yunnan, Sichuan and Guizhou Province. Its seeds contain 27–40% oil that can be processed to produce a high-quality biodiesel. It is estimated that there are about 36.7 thousand ha natural distribution area and 116.7 thousand ha planted area of *Jatropha curcas* in China. According to the output of 1 ton per hectare, there are 153.4 thousand seeds per year. At present, the seeds are mainly used for biopharmaceutical, biological pest control, ecological construction.

Helianthus tuberous L

Helianthus tuberous L is one of the 90 species of plants within the sunflower family. It is widely grown in both North and South China but only at scattered locations. The tubers are elongated and uneven, typically 7.5–10 cm long and 3–5 cm thick, and contain fructose polymer which can be processed into biodiesel.

Pistacia chinensis

Pistacia chinensis is a tall deciduous tree native to central and western China. It can be cultivated in harsh conditions and poor-quality soils. It was mainly distributed on the hilly and mountainous areas in more than 23 provinces in North, Central and South China. The seeds of *Pistacia chinensis* contain 25% of oil that can be refined into biodiesel.

Xanthoceras sorbifolia Bunge

Sorbifolia is an unique kind of deciduous shrub originated in the Loess Plateau. It can tolerant undesirable environmental conditions such as drought and low fertility, and distributes in North and Northeast China, such as Shanxi, Inner Mongolia, Hebei, Gansu. The seeds contain about 35–40% of oil, which can be used to produce biodiesel.

As the distribution of *Helianthus tuberous L, Pistacia chinensis* and *Xanthoceras sorbifolia Bunge* are scattered and relatively scarce, there are no official statistics on their planting area and yield. During the ecological construction in China these years, these plants were grown as ecological species in mountainous area. The tubers of *Helianthus tuberous L* are mainly used for food, starch or alcohol. The seed of *Pistacia chinensis* can be used to make soap, lubricants, medicine or edible oil. The seed of *Xanthoceras sorbifolia Bunge* is usually used for edible oil, and its leaves can be processed into tea. However, due to lower production, there are few large factories with these fruits as raw materials.

2.3   Potential of Marginal Land for Energy Plants

Each energy crop has its environmental requirements on climate, terrain and soil. Based on the experts' technical recommendation and related research (Zhuang et al. 2011; Wu et al. 2009; Wang et al. 2012; Zhang 2012; Lin et al. 2007), this study selected 6 factors to determine the spatial extents of marginal land suitable for energy plants, including average annual temperature, effective accumulated temperature (≥ 10 °C), annual precipitation, soil organic matter, soil texture and slope. Using the identified marginal land (Fig. 4) as a spatial mask, the spatial distribution of marginal land suitable for energy plants in China was identified based on the environmental requirements of each species (Table 1). The results are shown in Fig. 5. The results indicated that the total area of marginal land suitable for energy plants was estimated up to 30.85 million ha, including 27 million ha suitable land and 4.64 million ha moderate suitable land (Table 2).

2.4   Potential of Bioenergy Production from Energy Plants
on Marginal Land

Based on the area of marginal land suitable for energy plants along with their yields and energy conversion rates, bioenergy production potentials were calculated (see Table 2). The results show that the potential output of ethanol production from *Manihot esculenta* and *Helianthus tuberous L* is 24.45 million tons, biodiesel from the other three energy plants is 8.77 million tons. Compared with the annual national biofuels target in 2020 (100 million tons of bio-ethanol, 2 million tons of bio-diesel)

Table 1 Eco-environmental requirements of main energy plants in China

		Manihot esculenta	Jatropha curcas L	Helianthus tuberous L	Pistacia chinensis	Xanthoceras sorbifolia Bunge
Average annual temperature (°C)	Suitable	≥21	≥18	−2–8	10–16	4–12
	Moderate Suitable	18–21	14–20	8–12	6–10	0–4,12–18
	Unsuitable	<18	<14	<−2,≥12	<6,≥16	<4,≥18
≥10 °C effective accumulated temperature (°C)	Suitable	≥6500	≥6000	≥2500	≥3000	2000–4000
	Moderate Suitable	5500–6500	5000–6000	2000–2500	1180–3000	1000–2000, 4000–5000
	Unsuitable	<5500	<5000	<2000	<1180	<1000,≥5000
Annual precipitation (mm)	Suitable	1200–2000	900–2000	450–1000	700–1300	400–800
	Moderate Suitable	600–1200, 2000–2300	500–900, 2000–2300	160–450, 1000–2000	500–700, 1300–1600	150–400, 800–950
	Unsuitable	<600,≥2300	<500,≥2300	<500,≥2000	<500,≥1600	<150,≥950
Soil organic matter (%)	Suitable	≥2	≥2	≥2	≥2	≥1.5
	Moderate Suitable	0.6–2	0.6–2	0.6–2	0.6–2	0.3–1.5
	Unsuitable	<0.6	<0.6	<0.6	<0.6	<0.3
Soil texture (a. sand (1–0.05 mm) composition, b. Silt (0.05–0.01) composition, c. clay (<0.001) composition, %)	Suitable	a < 50 and c < 30	a < 50 and c < 30	a < 50 and c < 30	a < 50 and c < 30	a < 50 and c < 30
	Moderate Suitable	70 > a ≥ 50	70 > a ≥ 50 or c ≥ 30	70 > a ≥ 50 or c ≥ 30	70 > a ≥ 50 or 40 > c ≥ 30	70 > a ≥ 50 or 40 > c ≥ 30
	Unsuitable	a > 70 or c > 40	–	–	c > 40	c > 40
Slope (°)	Suitable	<10°	<10°	<10°	<10°	<5°
	Moderate Suitable	10–25°	10–25°	10–25°	10–25°	5–15°
	Unsuitable	≥25°	≥25°	≥25°	≥25°	≥15°

mandated by China's Medium- to Long-term Renewable Energy Development Plan, China has great potential to develop biofuel crops on marginal land.

3 Economic Feasibility of Using Marginal Land for Biofuels Production

Although China has sufficient marginal land to meet its biofuel target, the development of bioenergy industry is still constrained by various factors, such as the supply and cost of feedstock, technological barriers, biofuels' revenues, and public

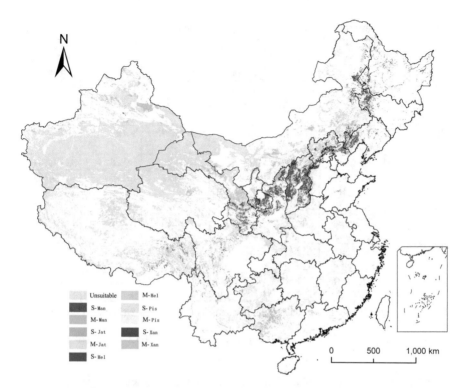

Fig. 5 Spatial distribution of marginal land suitable for energy plants in China (S-Suitable, M-Moderate Suitable; Man-*Manihot esculenta*, Jat-*Jatropha curcas L*, Pis-*Pistacia chinensis*, Hel-*Helianthus tuberous L*, Xan-*Xanthoceras sorbifolia Bunge*)

acceptance. The entire bioenergy production chain has two critical components: the production of bioenergy feedstocks by farmers and the industrialized biofuels production at biofuel refineries. On one hand, farmers' willingness to cultivate energy plants will affect the production and supply of feedstock to bioenergy refineries. On the other hand, the costs and revenues of bioenergy refineries are also a key factor for sustain bioenergy production. Therefore, this study analyzed the economic feasibility from the respect of farmers' net income and the biofuel refineries' net revenue.

3.1 Farmers' Input-Output of Growing Energy Plants

The economic revenue from energy plants dominates farmers' land use decision. Despite increasing price of fertilizers, seeds, pesticides, machinery and labor, prices of agricultural products have remained at a low level in recent years. Actually, low agricultural income has forced large quantity of rural labors to migrate to

Table 2 Bioenergy production potential of energy plants in China

		Land area (ha)	Yield per unit area (kg/ha)[b]	Total output (t)	Energy conversion rate[b]	Biofuel production potential (t)
Man[a]-ethanol	Suitable	27	16,000	432	1:0.130	56
	Moderate Suitable	4,637,879	12,000	55,654,548	1:0.130	7,235,091
Jat[a]-biodiesel	Moderate Suitable	5,949,627	1000	5,949,627	1:0.25	1,487,407
Pis[a]-biodiesel	Suitable	2,134,111	6000	12,804,666	1:0.25	3,201,167
	Moderate Suitable	1,669,703	4000	6,678,812	1:0.25	1,669,703
Hel[a]-ethanol	Suitable	2,039,813	50,000	101,990,650	1:0.063	6,425,411
	Moderate Suitable	7,415,601	30,000	222,468,030	1:0.063	14,015,486
Xan[a]-biodiesel	Suitable	2,127,776	1500	3,191,664	1:0.30	957,499
	Moderate Suitable	4,878,064	1000	4,878,064	1:0.30	1,463,419

Note: [a]Man-*Manihot esculenta*, Jat-*Jatropha curcas L*, Pis-*Pistacia chinensis*, Hel-*Helianthus tuberous L*, Xan-*Xanthoceras sorbifolia Bunge*.
[b]Data from Fang et al. (2010), Wu and Huang (2010), Xie et al. (2007), Liu et al. (2010)

urban areas. The crops cultivated on marginal land may present even less economic incentives for biomass crops adoption.

Based on our field survey and literature review, the input-output of *Manihot esculenta*, *Jatropha curcas L*, *Helianthus tuberous L* and *Xanthoceras sorbifolia Bunge* are shown in Tables 3 and 4. On average, the net revenues of producing annual crops *Manihot esculenta* and *Helianthus tuberous L* are 3639.75 yuan/ha and 36577.5 yuan/ha respectively if the land and labor costs are taken into account. The gross revenues are 10074.75 yuan/ha and 52327.5 yuan/ha respectively. Perennial crops, *Jatropha curcas L* and *Xanthoceras sorbifolia Bungein*, show negative net revenues during the first few years. Starting the 7th year, their net revenues shifts to 223 yuan/ha and 2370 yuan/ha respectively with gross revenues as 1255 yuan/ha and 5220 yuan/ha respectively.

The results indicated that the net revenue of energy plants' cultivation is relatively low. Especially with very low land availability (0.14 ha) per person in rural China, this prospect of extensive energy crop cultivation is dire. Considering the high transportation costs and opportunity cost of non-agricultural work, many families would rather abandon these arable land. In addition, the high uncertainty in bioenergy crops prices also eliminated the enthusiasm of farmers to cultivate energy plants. For example, the price of *Xanthoceras sorbifolia Bungein* peaked at 80 yuan/kg in Yuzhong, Gansu Province in 2011, but continuously declined to 20 yuan/kg in 2014 partially due to the over-supplies in the market.

Table 3 Input-output of *Manihot esculenta* in Guangxi and Helianthus tuberous L in Ningxia (Unit: yuan/ha)

	Seed	Fertilizers	Pesticide	Plastic film	Employment machinery	Labor input[c]	Land input[b]	Economic output[d]	Income (Include land and labor input)	Income (Not include land and labor input)
Manihot esculenta (Wuming, Guangxi, 2010)[a]	1053	3121.95	215.7	132	402.75	4935	1500	15000.22	3639.75	10074.82
Helianthus tuberous L (Yancheng, Jiangsu, 2014)[a]	2587.5	1560	–	–	2025	9750	6000	58,500	36577.5	52327.5

Note: [a]Data of Helianthus tuberous L were from Zhang et al. (2015); data of *Manihot esculenta* were from authors' field survey in Wuming county Guangxi, April, 2010
[b]Land cost refers to the lease price of the land in the same quality, which is 1500 yuan/ha in Guangxi in 2010, 6000 yuan/ha in Jiangsu in 2014
[c]Labor costs are associated with all on-farm activities, such as land preparation, seeding, irrigation, fertilizing, weeding, harvesting, transportation. The labor input cost refers to the price of male labor at local place that year. Labor price is 80 yuan/day in Guangxi 2010, 120 yuan/day in Jiangsu 2014
[d]The average production of *Manihot esculenta* in Guangxi 2010 was 28302.3 kg/ha, with the price of 0.53 yuan/kg, the Economic output was 15000.22 yuan/ha. The average production of *Helianthus tuberous L* in Jiangsu 2014 was 45,000 kg/ha, with the price of 1.3 yuan/kg, the Economic output was 58,500 yuan/ha

Table 4 Input-output of *Jatropha curcas L* in Gansu and *Xanthoceras sorbifolia Bungein* in Yunnan

	Input-output	1st year	2nd year	3rd year	4th year	5th year	6th year	7th year
Jatropha curcas L (5 counties in Yunnan, 2008)[a]	Seeds (yuan/ha)	583.00	0.00	0.00	0.00	0.00	0.00	0.00
	Fertilizers (yuan/ha)	183.00	231.00	113.00	22.00	24.00	28.00	28.00
	Labor input (yuan/ha)[c]	2294.00	1205.00	548.00	274.00	383.00	657.00	657.00
	Land input (yuan/ha)[b]	375.00	375.00	375.00	375.00	375.00	375.00	375.00
	Physical output (kg/ha)	0.00	120.00	270.00	465.00	645.00	855.00	855.00
	Economic output (yuan/ha)[d]	0.00	180.00	405.00	698.00	968.00	1283.00	1283.00
	Income (Include land and labor input) (yuan/ha)	**-3435.00**	**-1631.00**	**-631.00**	**27.00**	**186.00**	**223.00**	**223.00**
	Income (Not include land and labor input) (yuan/ha)	**-766.00**	**-51.00**	**292.00**	**676.00**	**944.00**	**1255.00**	**1255.00**
Xanthoceras sorbifolia Bungein (Yuzhong county in Gansu, 2015)[a]	Seeds (yuan/ha)	1110.00	0.00	0.00	0.00	0.00	0.00	0.00
	Fertilizers (yuan/ha)	1200.00	0.00	0.00	0.00	0.00	0.00	0.00
	Pesticide, plastic film (yuan/ha)	1440.00	0.00	0.00	0.00	0.00	0.00	0.00
	Labor input (yuan/ha)[c]	8250.00	1500.00	1500.00	1800.00	1800.00	1800.00	1950.00
	Land input (yuan/ha)[b]	900.00	900.00	900.00	900.00	900.00	900.00	900.00
	Physical output (kg/ha)	0	0	0	108.75	152.25	191.4	261
	Economic output (yuan/ha)[d]	0	0	0	2175	3045	3828	5220
	Income (Include land and labor input) (yuan/ha)	-12,900	-2400	-2400	-525	345	1128	2370
	Income (Not include land and labor input) (yuan/ha)	-3750	0	0	2175	3045	3828	5220

Note: [a]Data on Jatropha curcas L were from Wu and Huang (2010). The field survey was carried out in Yunnan, 2008. Data on *Xanthoceras sorbifolia Bungein* were from authors' field survey in Yuzhong County, August, 2015

[b]Land cost refers to the lease price of the land in same quality, which is 375 yuan/ha in Yunnan 2008, 900 yuan/ha in Gansu 2015

[c]Labor costs are associated with all on-farm activities, such as, land preparation, seeding, irrigation, fertilizing, weeding, harvesting, transportation. The labor input cost refers to the price of male labor at local place that year. Labor price in Yunnan in 2008 was 36.5 yuan/day; but it was 100 yuan/day in Gansu 2015

[d]The price of *Jatropha curcas L* seed was 1.5 yuan/kg in Yunnan in 2008. The price of *Xanthoceras sorbifolia Bungein* was 20 yuan/kg in Gansu in 2014

3.2   Bioenergy Plants' Input-Output of Producing Bioethanol and Biodiesel

At present, the most significant problem restricting the development of bioenergy plants in China is limited supplies and high cost of feedstock. Before these plants were identified as bioenergy feedstock, most of them were sporadically distributed in marginal land due to their low economic benefits. For example, *Manihot esculenta*, cultivated in some southern provinces, was only used for *Manihot esculenta* flour. Thus, farmers have no willingness to grow these crops.

However, after these crops are identified as bioenergy feedstock, the cultivation may be highly profitable at their early development stage due to high demands for seeds and nursery stock. Higher prices, albeit beneficial for farmer's income, become a critical prohibitive factor for large-scale production in bioenergy plants. For example, according to the conversion rate of 1 to 0.3 between *Xanthoceras sorbifolia* seeds and biodiesel, the production of one ton of biodiesel will need 3.33 tons of *Xanthoceras sorbifolia* seeds. Assuming profitable market price of *Xanthoceras sorbifolia* is 20 yuan/kg, the cost of producing one ton of biodiesel would be 66,600 yuan, much higher than the market price of biodiesel (4019 yuan/t). Sometimes, the current production scale of bio-feedstock could not meet the needs of energy plants. For example, Guangxi COFCO Bio-energy Co. Ltd., a Beihai-based bioenergy company that use *Manihot esculenta* to produce biofuels, need 1.4 million tons fresh *Manihot esculenta* to meet its 200,000 ton/year production capacity. However, most regional supplies of the *Manihot esculenta* were used for producing *Manihot esculenta* starch, which forced the company to import *Manihot esculenta* from Thailand, or even reduce production. Huge demands for *Manihot esculenta* and Thailand's restrictions on *Manihot esculenta* exports have led to rising *Manihot esculenta* prices. The increase in raw material costs also constrains the development of bioenergy plants.

Table 5 presented the input and price of bioenergy production per ton. The feedstock accounts for most of the total cost, which is much higher than biofuel prices. Since the price of bioenergy is strictly controlled by the government, the price of fuel ethanol is bundled with 90# gasoline price with the coefficient of 0.9111. If the price of 90# gasoline is 7000 yuan/t, the price of fuel ethanol is 6377.7 yuan/t. With the decline of international oil prices, the price of fuel ethanol and biodiesel declined correspondingly. It was reported that the price of biodiesel in late November 2015 was 4019 yuan/t. Therefore, the cost of feedstock was even higher than the sale price of bioenergy. In present, China's fuel ethanol production focuses on *Manihot esculenta* and maize, biodiesel production mainly relies on *Pistacia chinensis* and *Jatropha curcas L.* Almost all plants are facing insufficient supplies and high prices of feedstock, which resulted in a large number of bioenergy plants failures.

Table 5 Input and price of bioenergy production per unit

	Feedstock	Feedstock (t)	Price of feedstock (yuan/kg)	Cost of feedstock (yuan/t)	Average cost of production (yuan/t)[b]	Expenditure (yuan/t)[b]	Price of biofuel (yuan/t)[c]
Ethanol	Fresh Man[a]	7.69	0.55	4230	About 260	4490	6377.7
	Fresh Hel[a]	15.87	0.60	9522	About 120	9642	6377.7
Biodiesel	Jat[a]	4.00	1.00	4000	About 480	4480	4019
	Pis[a]	4.00	1.50	6000	About 500	6500	4019
	Xan[a]	3.33	20.00	66,600	About 500	67,100	4019

Note: [a]Man-*Manihot esculenta*, Jat-*Jatropha curcas L*, Pis-*Pistacia chinensis*, Hel-*Helianthus tuberous L*, Xan-*Xanthoceras sorbifolia Bunge*

[b]The net income is only a rough estimate data. With regard to expenditure, because of the unavailability of data, this paper only considered the cost of feedstock and auxiliary materials. There are many other expenses during bioenergy production, such as staff salaries, machine depreciation, finance, management, sales, transportation and other expenses. As the income, there are a part of by-products during production process generally, which can increase part of the income

[c]The prices of ethanol and biodiesel are constantly changing. We chose the about prices in late 2015 just as an example in this table

3.3 The Impacts of Government Regulations in Bioenergy Development

In China, bioenergy production was mainly dependent on grain crops. The development of bioenergy in China was affected by the government regulations significantly (Chen et al. 2015). Policies and bioenergy subsidies directly affected the development of bioenergy feedstock and industry. From 2001 to 2005, in order to reduce the stale maize and wheat from the state grain stocks, the government sanctioned a series of policies and legislations to promote bioenergy production. During this period, the government subsided bioenergy plants in a variety of ways, such as low price of feedstock, tax subsidies. For example, the stale wheat was supplied to Tianguan Group Co. Ltd., Henan at 1.14 yuan/kg, lower than the market price of new wheat 1.44 yuan at that time. These companies were also exempted from a 5% consumption tax. Value added tax (normally 17%) was levied and then refunded at the end of each year. These subsidies helped the profits of the plants.

From 2006 to 2008, a fixed level of direct subsidy was offered by the central government to ensure profits for each bioethanol plant. The subsidy was 1883 yuan/t in 2005, 1628 yuan/t in 2006, 1373 yuan/t for both 2007 and 2008. Apart from the subsidies in the production stage, two state-owned petroleum companies, China National Petroleum Corporation (CNPC) and China Petroleum and Chemical Corporation (CPCC), are compulsorily required to sell bioethanol-blended gasoline in sale stage. However, this policy only applies to the big-four state-owned plants. Private plants are not allowed to enter the E10 market. In addition, during this period, with the depletion of stale grain, some bioenergy plants used new grains as feedstocks. In order to ensure food security, the government proposed to develop renewable energy from non-food crops. In 2006, COFCO Bioenergy Co., Ltd. which used *Manihot esculenta* as feedstock was established in Guangxi, with the capacity of 200,000 tons per year.

Since 2009, the bioenergy subsidies shifted from a fixed scheme to a dynamic one. Under this scheme, the government will not issue the subsidy to the plants until the revenues of biofuel sales are below the production costs. The subsidy was mandated at 500 yuan/t and 750 yuan/t to bioethanol that was produced from grains and non-grain feedstocks respectively in 2012. With the government no longer supporting the grain bioethanol, the subsidy for grain biofuel production decreased from 300 yuan/t in 2013 to 200 yuan/t in 2014, and 100 yuan/t in 2015, none in 2016. Since the subsidy to non-grain energy plants still stayed at 750 yuan/t, some grain energy plants transitioned to non-grain energy plants. For example, the COFCO changed the 135,000 t/year of corn ethanol production equipment into *Manihot esculenta* ethanol in 2013. The non-grain bioenergy production is likely to increase in the next few years if current subsidies for non-grain bioenergy remain stable.

4   Dilemma and Future Development of Biofuels in China

Although China's non-grain bioenergy production has developed fast during the last few years, it still faces severe challenges. One of the most important constraints is the feedstock supply (Wang 2011). To meet the large amount of feedstock for bioenergy plants' operation, there should be large-sale cultivation and production of non-grain energy plants. However, to avoid biofuel feedstock competition with food, biofuel crops must be grown on marginal land in China. Constrained by the water-shortage and barren soil, the input-output ratio of energy plants in marginal land is low. Coupled with large rural labors transferred to cities, the remaining labors dominated by the seniors have less incentive to grow bioenergy plants.

Meanwhile, bioenergy production on marginal land faced a vicious circle. Low input-output ratio of bioenergy plants leads to farmers' little incentive for planting bioenergy plants. As the lack of feedstock supply, many plants are reluctant to invest in the non-grain bioenergy production. This further constrain the planting of bioenergy plants. We found that it was almost infeasible to develop bioenergy based on marginal land if only rely on the market.

At present, China's bioenergy plants' planting and biofuel production both depend on government subsidies (Ji and Long 2016). In order to achieve sustainable biofuel crops production on marginal land, the biofuel crops growers are expected to own the following characteristics: heavily initial investment, advanced technology, strong financial position, and high resilience to market risk. Taking these factors into account, large companies are more likely to succeed than individual farmers.

As rural labors increasingly migrated to urban areas, leasing land is common in China. Especially in western China, companies can rent board marginal land at relatively low prices. For example in Yuzhong, Gansu, the leasing cost only 900 yuan/ha in 2015. Some companies have rent dozens or even hundreds hectares of wasteland for the cultivation of *Xanthoceras sorbifolia Bungein*. It can be expected that large-scale production and establishment of bioenergy plants occur in the next few years.

5   Conclusion

Using marginal land for bioenergy production was considered an important strategy to promote this clean energy in China. Based on reasonable definition, the latest official land use data, and field survey information, we found:

1. The total area of marginal land in China was 289.71 million ha. Among it, the total area of marginal land that can be used for cultivating five major species of energy plants was about 30.85 million ha, including 27 ha suitable and 4.64 million ha moderate suitable for *Manihot esculenta*, 5.95 million ha moderate suitable for *Jatropha curcas L*, 2.04 million ha suitable and 7.42 million ha moderate suitable for *Helianthus tuberous L*, 2.13 million ha suitable and 1.67

million ha moderate suitable for *Pistacia chinensis*, 2.13 million ha suitable and 4.88 million ha moderate suitable for *Xanthoceras sorbifolia Bunge*.

2. Assuming that these five energy plants are well adapted on the marginal land, the potential output of ethanol production from *Manihot esculenta* and *helianthus tuberous L* would be 24.45 million tons, the biodiesel would be 8.77 million tons from *Jatropha curcas L*, *Pistacia chinensis* and *Xanthoceras sorbifolia Bunge*.

3. Although China has sufficient marginal land for energy plants' cultivation, bioenergy industry faces challenges due to economic feasibility. From the farmers' perspective, low revenues led to little incentive of energy plants' cultivation. From the bioenergy producers' perspective, unstable supply and high cost of feedstock constrained the normal production. In China, both energy crops' cultivation and bioenergy production heavily replies on government subsidies. It was infeasible to develop biofuels suing marginal land by solely relying on the market at present.

In order to foster a healthy bioenergy production chain, a balanced incentive scheme is needed. The government subsidies should not only focus on biofuel production, but also benefit the cultivation of feedstock. Technological and other financial supports for new and existing farmers as well as enterprises to grow energy plants on marginal land should be in place. With the scenarios of large area of marginal land converted for energy plants' cultivation, the ecological and environmental implications associated with the bioenergy development should be addressed with more research.

Acknowledgment The study is funded by the ministry of land and resources public welfare scientific research special project (No. 201311125) and the National Natural Science Foundation of China (No. 41201567).

References

Chang S, Zhao L, Govinda TR et al (2012) Biofuels development in China: technology options and policies needed to meet the 2020 target. Energy Policy 51:64–79

Chen H, Xu M, Guo Q et al (2015) A review on present situation and development of biofuels in China. J Energy Inst. https://doi.org/10.1016/j.joei.2015.01.022

Chen W, Wu F, Zhang J (2016) Potential production of non-food biofuels in China. Renew Energy 85:939–944

Fang J, Pu W, Zhang H (2010) The development status of Cassava industry ant home and abroad. Chinese Agric Sci Bull 26(16):353–361. (in Chinese)

FAOSTAT (2014) [Online] http://www.liuxue86.com/a/2706001.html. http://www.fao.org/faostat/en/#data/QC

Fu C, Wu F (2014) The estimation of producing potential of fuel ethanol and its developing strategy in China. J Nat Resour 29(8):1430–1440. (in Chinese)

Gelfand I, Sahajpal R, Zhang X et al (2013) Sustainable bioenergy production from marginal lands in the US Midwest. Nature 493(7433):514–517

Ji X, Long X (2016) A review of the ecological and socioeconomic effects of biofuel and energy policy recommendations. Renew Sust Energ Rev 61:41–52

Kang S, Wilfred PM, Wang D et al (2013) Hierarchical marginal land assessment for land use planning. Land Use Policy 30:106–113

Li Z (2008) The development, problem and policy suggestions of bio-ethanol. China Biotechnol 28(7):139–142

Lin C, Li Y, Liu J et al (2007) A study on the diversity and exploiting prospect of energy plant resources. J Henan Agric Sci 36(12):17–21

Liu G, Liu Y, Huang C et al (2010) Phychemical properties and preparation of bio-diesel with five categories of woody plant seeds oil. Acta Agric Univ Jiangxiensis 32(2):0339–0344. (in Chinese)

Qiu H, Huang J (2008) Global bio-energy development and its impact on the prices of agricultural products [J]. World Environ 4:19–21. (In Chinese)

Swinton SM, Babcock BA, James LK et al (2011) Higher US crop prices trigger little area expansion so marginal land for biofuel crops is limited. Energy Policy 39:5254–5258

Tang Y, Xie J, Geng S (2010) Marginal land-based biomass energy production in China. J Integr Plant Biol 52(1):112–121

Wang Q (2011) Time for commercializing non-food biofuel in China. Renew Sust Energ Rev 15:621–629

Wang F, Shi X (2015) Geospatial analysis for utilizing the marginal land in regional biofuel industry: a case study in Guangdong Province, China. Biomass Bioenergy 83:302–310

Wang L, Yang H, Feng Z et al (2012) Potential capacity of cassava planting in Guangxi: assessment of natural suitability and socio-economic restrictions. Resour Sci 34(1):150–158

Wu W, Huang J (2010) Economic feasibility analysis of Jatropha Curcas plant for biodiesel production. China Rural Econ 7:56–63. (In Chinese)

Wu W, Huang J, Deng X (2009) Potential land for plantation of Jatropha curcas as feedstocks for biodiesel in China. Sci China Ser D-Earth Sci. https://doi.org/10.1007/s11430-009-0204-y. (in Chinese)

Xie G, Guo X, Wang X et al (2007) An overview and perspectives of energy crop resources. Resour Sci 29(5):74–80. (in Chinese)

Zhang X (2012) The growth adaptability of Xanthoceras Srbifolia Bge of different seed source region and its salt tolerance in Shendong Mining Area. Master thesis of Inner Mongolia Agricultural University. 05. (in Chinese)

Zhang Q, Ma J, Qiu G et al (2012) Potential energy production from algae on marginal land in China. Bioresour Technol 109:252–260

Zhang Z, Zhang Y, Luo J et al (2015) Economic benefit analysis of industrialized planting of Helianthus tuberous L in Jiangsu Coastal Region. J Jiangsu Agric Sci 43(8):480–483. (in Chinese)

Zhao L, Chang S, Wang H et al (2015) Long-term projections of liquid biofuels in China: uncertainties and potential benefits. Energy 83:37–54

Zhuang D, Jiang D, Liu L et al (2011) Assessment of bioenergy potential on marginal land in China. Renew Sust Energ Rev 15:1050–1056

Production of Energy Crops in Heavy Metals Contaminated Land: Opportunities and Risks

Bruno Barbosa, Jorge Costa, and Ana Luisa Fernando

Abstract An increasing global awareness that the supply and security of petroleum-based materials is diminishing, coupled with environmental concerns related to climate change, water availability, and soil degradation, has increased demand for more renewable, diversified, and sustainable systems, of which biomass resources are one of the pillars. Yet, the demand for biomasses may increase sharply, thus increasing the risk of conflicts on land use due to competition for food and feed. Hence, segregating the growth of dedicated biomass crops on contaminated land is considered a suitable option to overcome these conflicts. In fact, most of energy crops are considered tolerant to soil contamination, and the cultivation of those crops can be considered an approach to restore or attenuate and stabilize contaminated soils while bringing additional revenue to owners. But, is it sustainable to produce energy crops in contaminated land? Yields and biomass quality can be affected by the contamination, reducing the energy and the greenhouse savings and compromising its economic exploitation. Nonetheless the production of energy crops on contaminated land may contribute also to improve the quality of soil and the biological and landscape diversity. In this context, studies on the production of energy crops in heavy metals contaminated land are reviewed, taking into account environmental, economic and socio-economic aspects. In the end, a critical assessment of the literature is made and opportunities and risks are pointed out.

Keywords Energy crops · Bioenergy · Land use change · Heavy metals contaminated land · Phytoremediation · Sustainability

B. Barbosa
Universidade Nova de Lisboa, Faculdade de Ciências e Tecnologia, Departamento de Ciências e Tecnologia da Biomassa, MEtRiCS, Caparica, Portugal

Universidade de São Paulo, São Paulo, Brazil

J. Costa · A. L. Fernando (✉)
Universidade Nova de Lisboa, Faculdade de Ciências e Tecnologia, Departamento de Ciências e Tecnologia da Biomassa, MEtRiCS, Caparica, Portugal
e-mail: ala@fct.unl.pt

© Springer International Publishing AG, part of Springer Nature 2018
R. Li, A. Monti (eds.), *Land Allocation for Biomass Crops*,
https://doi.org/10.1007/978-3-319-74536-7_5

1   Introduction

Nowadays, environmental issues are in the spotlight, stimulating discussions at several levels: technological, economic, political and social. The need to provide sufficient available primary energy sources and secondary energy forms coupled with the high exploitation in its use by an exponentially growing population, is a global issue and receiving attention towards its sustainability (environmental, economic and socio-economic). In fact, most of the energy supply is based on fossil resources, which are finite, non-renewable and its use generates greenhouse gas (GHG) emissions and other pollutants, such as sulfur dioxide, volatile organic compounds and heavy metals (Nicoletti et al. 2015). Therefore, the use of alternative forms of energy production is crucial and a transition to a more diversified energy system is compulsory.

Energy crops, annual or perennial, arise as a viable substitute to non-renewable energy sources. Biomass crops are renewable feedstocks, since through the process of photosynthesis, sunlight energy, CO_2 and H_2O are fixed and stored in plants being converted into bioenergy in a short timescale. Additionally, bioproducts are biodegradable and compatible with the existing systems. Perennial crops, such as switchgrass (*Panicum virgatum* L.) or giant reed (*Arundo donax* L.), are highly productive and low-maintenance crops that can be used for heat and second generation biofuels production (Balan et al. 2012; Fernando et al. 2016b). Annual crops, such as the oil crop camelina (*Camelina sativa* (L.) Crantz.), or the fiber crop kenaf (*Hibiscus cannabinus* L.), are multipurpose crops that can provide valuable feedstocks for the production of bio-based commodities (bioplastics, lubricants, composites) and bioenergy (Berti et al. 2016; Lips and Van Dam 2013). But, the increased commitment of cropland to energy crops increases the competition for agricultural land, which might boost food commodities prices, accentuating the fuel versus food dilemma, and the land use change debate (Dauber et al. 2012). Furthermore, to meet the biomass procurement, GHG emission benefits might be offset owing to the direct and indirect disturbance of carbon storage in the soil and vegetation, and the intensiveness of the cultivation and the increased pressure on natural resources may jeopardize the bio-systems environmental sustainability.

Land lies in the intersection of these issues and its use is a key point in the overall sustainability of bioenergy systems. The European Commission has repeatedly declared the intention to avoid the cultivation of non-food, and especially energy crops in fertile agricultural land, in order to avoid the consequential effects in the food market (EC 2009; EC/JRC 2013). Therefore, it is reasonable to encourage agricultural land use planning to avoid competition with food production. Interest has risen in the cultivation of unused lands, specifically sites degraded by anthropogenic interventions. Likewise, areas fouled by landfills, mining and industrial activities or polluted sediment depositions might be occupied with energy crops tolerant to the contamination (Bosco et al. 2016; Schröder et al. 2008). This approach is a promising option to sustain and improve rural development, especially in areas threatened by abandonment, helping to restore or attenuate and stabilize contaminated sites while bringing additional revenue to owners. Additionally, land use competition with food crops and land use change debates are minimized. Tolerance to contaminated soils and phytoremediation capabilities have been suggested for perennial grasses, such as *Miscanthus* spp. and giant reed (Barbosa et al. 2015a; Nsanganwimana et al. 2014a, b), short rotation coppices, such as poplar trees (Redovniković et al. 2017) and annual crops, such as hemp, flax and kenaf (Pandey et al. 2016) among other energy crops.

Therefore, the aim of this work is to provide a comprehensive review of studies on the production of energy crops in contaminated land in order to identify opportunities and risks associated with it. Focus was on soils contaminated with heavy metals once those contaminants belong to the most frequent in polluted soils.

2 Soils Contaminated with Heavy Metals: Soil-Metal-Plant Relationships

Soil contamination is a widespread problem in Europe. The most frequent contaminants are heavy metals and mineral oil, and approximately three million sites are estimated to have been potentially affected by activities that can pollute soil. From those, approximately 250,000 may need urgent remediation (UWE 2013).

Heavy metals occur as natural elements in the earth crust and both natural geological processes and human activities contribute to their dissemination in the environment (Benjamin and Honeyman 1992). Anthropogenic activities, such as use of pesticides, fertilizers, fossil fuels, mining, emissions from steel, metallurgical and petrochemical industries, are increasing the rate of release of heavy metals, contributing to the contamination of soils, where they tend to deposit (Kabata-Pendias 2011; Wuana and Okieimen 2011). The disposal of materials containing heavy metals – a long list which includes paint, electronic waste, and sewage – also contributes to the burden of heavy metal contamination. Examples of areas with high occurrence of local contamination with heavy metals include the North-West Europe from Nord-Pas de Calais (Sabra et al. 2011), the Rhein-Ruhr region and the Saar region in Germany (Ginneken et al. 2007; Hüffmeyer et al. 2009), and Belgium and the Netherlands (Meers et al. 2005). Other contaminated areas comprise northern Italy (Bini et al. 2012; Giuseppe et al. 2014), Iberian Peninsula (Clemente et al. 2005; Costa and Jesus-Rydin 2001; Pratas et al. 2013), Poland (Herpin et al. 1996), Czech Republic and Slovakia (Ginneken et al. 2007) as well as Ukraine (Shcheglov et al. 2014; Vystavna et al. 2014), Bulgaria (Yordanova et al. 2014) and Greece (Papazoglou et al. 2005).

Heavy metals are mainly retained in the soil by the following processes (Baird 1999): (a) adsorption on the surface of the mineral particles; (b) complexation by humic substances; (c) precipitation. Contamination by heavy metals can seriously affect soil's capacity to accomplish its functions, especially when contamination exceeds a critical threshold. Contamination affect how much water soil can hold, the living organisms it supports, which chemical reactions are likely to occur, and how it cycles nutrients. The result can be a loss in organic matter and structure, and of nutrients due to the soil's lower holding capacity (Russel and Alberti 1998; Fernando 2005) with detrimental effects on the soil properties (fertility, organic matter, structure), reducing crop yields and quality of agricultural products. Furthermore, soil contamination involves several risks, namely effects in human and animal health, plant species and environment: DNA damage in animals and humans, implications for the food chain and ecosystem functioning, fodder crop contamination, food contamination, surface and groundwater contamination, etc. (Barbosa et al. 2016). Thresholds for most heavy metals exist in most countries, but variable, depending on the use of the soil (for agriculture, industry, or housing) and on the natural background concentrations. In fact, the background values of the various elements present in soils vary significantly from country to country, and even from region to region, making it difficult to adopt international limit values for metals in soils.

Heavy metal contaminated soils can be toxic to plants. The most common symptoms of heavy metal phytotoxicity are leaf chlorosis, growth inhibition, distur- bance of respiration and nitrogen metabolism, and the reduction of photosynthesis and water and nutrient uptake (Wang et al. 2008). Heavy metal also generates oxidative stress which can lead to cell death. Heavy metal toxicity is dependent on the element, their concentration in soil, their bioavailability, soil pH, type of soil and on the plant growth stage and exposure time (Prasad 2004; Zárubová et al. 2015). Plant tolerance to heavy metals contamination depends on a large number of factors and varies significantly among plant species (Evangelou et al. 2013). Several physiological mechanisms may explain heavy metals tolerance by different plants. The plants may be protected externally against the metal, or else tolerate high concentrations of metals in their tissues, through specific physiological mechanisms that lead to its phytotoxicity minimization (Barbosa et al. 2016). For example, direct ionic metal toxicity can be quenched by anatomical changes, such as development of secondary sheath bundles and tissue scarification (Guo and Miao 2010) or by the plant ability to immobilize ions in the hypogeal components (Kabata-Pendias 2011). Another survival strategy is linked with the mycorrhizal association in the dense belowground organs, which provides a metal excluder barrier (Leung et al. 2007). Plants can also phytostabilize the contaminants in the soil, preventing its percolation or runoff to ground and surface waters. Plants may adsorb the toxic compounds present in the soil, in the root system, or may release organic compounds (exudates) to the soil by the root system, which can complex or precipitate the contaminants thus stabilizing/immobilizing the contaminants (Fernando 2005). Exudates can also stimulate microbial activity accelerating the degradation of organic contaminants to non-toxic forms or less toxic forms (Mirza et al. 2011).

Plants tolerant to soils contaminated with heavy metals can be classified as excluders, where metal concentrations in the shoots are maintained up to a critical value at a low level across a wide range of metal concentrations in soil, indicators, where internal concentrations in the plant reflect the soil concentrations, and accumulators, where metals are concentrated in aboveground plant fractions from low to high soil concentrations (Leung et al. 2007). Hyperaccumulators (e.g. *Brassica juncea* and *Thlaspi caerulescens*) are plants capable of growing in soils with very high concentrations of metals, and capable to uptake heavy metals at high rates through their roots, and concentrating extremely high levels of metals in their tissues (Fernando 2005). Hyperaccumulator plants from natural populations are plants that typically contain more than 1000 mg.kg^{-1} (0.1% of dry weight) of the less mobilized elements (Co, Cu, Cr, Pb and Ni); and more than 10,000 mg.kg^{-1} (1% of dry weight) of the easily mobilized elements (e.g. Mn or Zn), in its biomass (Prasad 2004). The high mobility of some elements from the belowground organs to the aboveground organs can be associated with their bounding to light organic compounds in xylem fluids (Kabata-Pendias 2011). Others, such as Cr or Pb, are less mobile because they bind to cell walls as pyrophosphates (Kabata-Pendias 2011).

This ability of hyperaccumulators to absorb, translocate and accumulate metals in their belowground and aerial components can be used to remove metals from contaminated media, a process currently known as phytoextraction (Fernando 2005). Nonetheless, hyperaccumulator productivities are usually low, and the resulting biomass is generally not economically attractive. In fact, model crops for heavy metals soil remediation should display: (a) fast growth and high yields; (b) extensive and deep root system and tolerance and ability to accumulate/degrade contaminants; (c) high efficiencies in the use of resources, high tolerance to pests, high adaptive capacity to soil and climatic conditions, as well as low ecological requirements; and (d) easiness to harvest and known agronomic techniques (Barbosa et al. 2016; Yang et al. 2005). Consequently, the use of crops, such as energy crops, can be a viable approach to add value to contaminated soils, once significant yields can be obtained and the biomass being produced can be economically valorized.

3 Opportunities Associated with the Production of Energy Crops in Contaminated Land

Establishing vegetation on contaminated land contributes to the capture of CO_2. Therefore, GHG emissions balance can represent a positive aspect when comparing with the nude-degraded land. Moreover, use of contaminated land for the production of biomass for heat, power, biofuels and biobased materials, can avoid further conflicts on land use competition for food and feed.

Benefits from the introduction of a vegetative cover in a contaminated land are linked with ecosystem services and soil function restoration (Chiaramonti et al. 2000; Cortina et al. 2011). The presence of vegetation and the incorporation of crops residues into soils may restore soil properties (fertility, structure, organic matter), control soil erosion and increase biological and landscape diversity (Fernando et al. 2014; Stewart et al. 2015). In fact, energy crops cultivation offers several environmental advantages and provides a wide range of ecosystem services.

Soil quality is affected by crops characteristics and crop management activities, which influences nutrient status, organic matter, pH, structure and erosion. The majority of energy crops present limited input-requirements, namely fertilizers, pesticides and water (Fernando et al. 2010a, b) and therefore it is claimed that common cropping management activities do not affect soil quality through the nutrient status. Yet, when comparing with bare contaminated land, all crops, more or less, disturb the soil's nutrient status (Fernando et al. 2016c, 2018). Therefore, fertilizer application should be as balanced as possible in order to avoid excessive deficit or surplus.

The vegetative cover and the rooting system provide soil structural enhancement and contributes to store carbon, improving soil organic matter accumulation (Brandão et al. 2010; Fernando et al. 2018). Perennial crops are reported to accumulate higher organic matter and to provide higher structural enhancement than annuals due to the less intensive soil amendment, to higher permanence in the soil, high inputs of residues and vigorous root development (Fernando et al. 2010a, b, 2018). In the study of Fernando (2013), long-term *Miscanthus* plantations were analysed and results showed that the presence of the crops in the soil contributed to an increment of organic matter by comparison with soils without the crop. Recycled carbon to soil from litter of *Miscanthus* represented *ca.* 3.1–3.9 Mg.ha^{-1}.year^{-1}. Carbon sequestration by the root and rhizome system was considerable, representing *ca.* 12.5–13.5 Mg.ha^{-1}, over the life time of the system. But this soil-rhizome accumulation may represent a concern, when the rhizomatous biomass has to be removed, due to changes of the land use, by release of the stored carbon. Other studies with perennial crops, also point out to the benefits of carbon storage under perennial crops fields (Monti and Zegada-Lizarazu 2016; Zatta et al. 2014). Annual crops with deep roots (hemp, sweet sorghum and sunflower) also increments soil organic matter and improves its structure in deeper horizons, with benefits for soil quality (Fernando et al. 2010b). In addition, accumulated organic carbon in the soil represents an option for carbon credit programs. Growing annual energy crops in crop rotation systems may also add benefits to the contaminated soil ecosystem. The insertion of annual energy crops in a rotation system can increase the yield and profitability of soil over time; promote disease and pest control on site by increasing biological and landscape diversity, as well as providing a stable source of biomass for fiber, bioenergy and other byproducts (Zegada-Lizarazu and Monti 2011).

Soils stabilization and control of soil erosion are also enhanced by the presence of energy crops as it was observed by Fernando et al. (2015, 2018) in their studies on the assessment of the environmental impact of perennial crops cultivation on less productive soils in the Mediterranean region. Cosentino et al. (2015), in their work in a sloppy area of Sicily, also observed that perennial crops present low erodibility potential and runoff. Although the above studies have not been carried out on contaminated soils, they show that the vegetative cover represent a mechanism to prevent soil erosion and the displacement of contaminants, through wind and water, thus protecting nearby terrestrial and aquatic ecosystems (Fernando et al. 2011).

Halting soil degradation and contamination with cultivation of energy crops has been argued as a promising solution as it was mentioned in several works, e.g. Pidlisnyuk et al. (2014). Indeed, the use of plants and their associated microbes, for soil decontamination, is a cost-effective, solar-driven technology that can complement or substitute physico-chemical approaches (Barbosa et al. 2016). Furthermore, as soil contamination involves several risks, as mentioned before in this chapter, the decontamination process will also contribute to ameliorate those effects.

Energy crops exhibit the required characteristics for remediation purposes and have demonstrated the ability to stabilize or lower the contaminant level of contamination in polluted soils. Willow and poplar have been documented as efficient landfill caps (Börjesson 1999; Duggan 2005) and in the treatment of soils contaminated with non-hazardous levels of industrial waste (Giachetti and Sebastiani 2006). Perennial herbaceous crops, such as giant reed, miscanthus and switchgrass are also considered suitable for soils contaminated with heavy metals (Arora et al. 2016; Barbosa et al. 2015a; Fernando and Oliveira 2004). Concerning annual crops, rape seed is documented for extracting heavy metals from contaminated soils (Sheng et al. 2008; Niu et al. (2007) successfully used the oilseed crops sunflower and Ethiopian mustard for uptaken metals from sewage sludge. Bioremediation capabilities have also been suggested for hemp (Linger et al. 2002), flax (Bjelková et al. 2001; Grabowska and Baraniecki 1997), kenaf (Catroga et al. 2005) and sweet sorghum (Epelde et al. 2009), among other crops.

Even though use of energy crops for remediation purposes is a possibility to clean contaminated sites, the decontamination response is usually very slow. According to several authors, it may be needed decades to effectively reduce the metal contamination through metal uptake by the aerial fraction of crops – the fraction that will be harvested and taken out from the field (Fernando 2005; Pidlisnyuk et al. 2014). This may represent a limitation of this technology application, especially when time is a key factor on choosing a solution for contamination removal. On the other hand, most of the works on the use of energy crops in contaminated soils, e.g. with heavy metals, indicate that the majority of contamination remain in the below ground biomass (e.g. Barbosa et al. 2015a). This mechanism of tolerance presented by energy crops can be beneficial, once the biomass quality remains similar, thus permitting the valuable use of the cropped biomass. When the level of contamination is high, yields can be reduced and, when contamination is with heavy metals, and they are in a more mobilized form, translocation to the aerial fraction of the biomass can occur. Translocation of the accumulated metals to the aerial fractions is usually displayed by the most mobilized metals (e.g. Zn and Cd). Regarding other metals, e.g. Pb and Cr, the translocation is verified for some contamination levels, but accumulating mostly in the leaves fraction and not in the stems (Barbosa et al. 2013; Fernando and Oliveira 2004; Sidella et al. 2013). To remediate contaminated soils, phytoextraction, a subprocess of phytoremediation in which plants remove metals from soil via accumulation in the biomass and its subsequent harvest, is considered the optimal approach (Ghosh and Singh 2005). Contaminants are retained by plant roots, with subsequent translocation to the aerial parts, followed by harvest and disposal/treatment of plant biomass (Mirza et al. 2011). According to Ghosh and Singh (2005) phytoextraction strategies are viable technologies to be applied in soils presenting diffuse contamination as well as in sites where their concentration is low and mainly located in soils surface. Concerning the phytoextraction process with energy crops, several studies, such as the ones of Hammer et al. (2003) and Meers et al. (2007), reported this ability. Hammer et al. (2003) tested the heavy metals (Cd and Zn) phytoextraction in two types of soils, limestone and acidic, which were contaminated, with *Salix viminalis*. The plant showed best results in acidic soil, presenting a higher biomass production and higher concentrations of metals in its aerial parts, especially in the leaves fraction. The application of iron in the chelate form, enhanced the production of biomass, but the same effect was not observed when sulphur was added. Meers et al. (2007) tested the phytoextraction with five Salix clones, in Cd, Cr, Cu, Ni, Pb and Zn contaminated soils and results showed a higher accumulation in Zn and Cd, in the aerial fraction, than the remaining tested elements. Yet, although the accumulation potential acts as beneficial when the purpose is oriented to clean the soils, the phytoaccumulation potential by the crops can also be detrimental. In fact, the accumulation of contaminants by the aerial biomass might cause a depreciation on the biomass quality for processing, limiting its industrial use.

Energy crops yields from soils contaminated with heavy metals can be stimulated by several mechanisms. One option is to stimulate the microbial community in the rhizosphere. In the work of Arora et al. (2016), switchgrass when inoculated with plant associated microbes (Arbuscular Mycorrhizal fungi and *Azospirillum*) showed enhanced metal absorption, increased the root length, branches, surface area, and root and shoot biomass, resulting in approximately 2-fold extraction of metals from Pb/Cd contaminated soils. Rajkumar and Freitas (2008) verified that under heavy metal contaminated soils, the inoculation with *Pseudomonas* spp. of *Ricinus communis* contributed to increase the aerial and root biomass growth and productivity, and the uptake of Zn (but not of Ni and Cu) by the roots, stems and leaves. These microbes produce organic acids, siderophores, and others which act as chelating agents, and form complexes with metals, decreasing their mobility in the soils, limiting the toxic effects of heavy metals and enhancing the biomass production in plants. Additionally, metal availability and toxicity to the host plant is reduced by means of metal absorption in mycorrhizal vacuoles and cell wall components, and by precipitation of metal oxalates in the intercellular spaces of the fungi (Arora et al. 2016). The addition of complexing agents, as ethylenediaminetetraacetic acid (EDTA) or citric acid, that can increase metals bioavailability and subsequent phytoextraction/ immobilization in plant tissues, can also enhance the growth and the tolerance of energy crops in heavy metals contaminated soils (Fernando 2005). Tests with *Sorghum bicolor*, sunflower (*Helianthus annus* L.) and willow showed that the addition of EDTA contributing significantly to increase the bioavailability of heavy metals in soils, which was reflected in the amount of metals accumulated in the biomass, enhanced by the increased yields of the biomass (Barbosa et al. 2016).

Decontamination of soils can also be accomplished by phytovolatilization, the use of plants to take up contaminants from the soil converting them into volatile forms or other less toxic compounds and transpiring them into the atmosphere (Ghosh and Singh 2005). This technology was primarily used in the removal of mercury, in the organometallic form, that was transformed into Hg-element, less toxic (Ghosh and Singh 2005). This technique was used with success in removing mercury (Hg) and arsenic (Mirza et al. 2011).

When the obtained biomass is processed, for fiber, energy and other bioproducts, and its exploitation is viable, the overall concept is fitting the biorefinery logics, and the circular economy.

Lewandowski et al. (2006) studied the economic value of the production of willow (*Salix* spp.) in Cd contaminated soils. Results were promising for both producers and authorities, but the economic value is dependent on factors, such as the end products, soil and climatic conditions, local financial situation and the dynamics of local markets, the time required to remediate the soil, and costs of soil remediation, which depend on crops production costs, with additional costs associated with the treatment of heavy metals contained in the biomass.

The phytoextraction technology shows also an additional interest, related to the possibility of metal recovery from the biomass after harvest (Mirza et al. 2011). Several approaches can be applied in order to recover heavy metals and other inorganic contaminants from biomass, and, at the same time, obtaining a product with economic value. Biomass can be composted, a process that will reduce the volume of the biomass, with added benefits towards the logistics (storage, transportation). Following, metals can be obtained from the composted biomass through leaching processes, which enhance metal solubility. However, thermochemical conversion processes are the main potential paths for converting biomass harvested from contaminated soils with heavy metals, with generation of electrical and thermal energy. When combusted, the process should be handled under controlled conditions in order to avoid gas and particles released into the atmosphere and the resultant ash should be disposed properly. If gasified, the metals oxides released from biomass should remain in the slag and the metal containing dust should be capture through techniques, such as modern flue gas clearing. If biomass is subjected to pyrolysis, special care should be taken regarding the coke, the pyrolysis residue containing heavy metals (Ghosh and Singh 2005). When the concentrations of the contaminants exceed limits imposed by regulations (e.g. European), the biomass or the resulting wastes should be appropriately disposed following countries rules.

In the study of Soldatos (2015), about the economics of cultivating perennial grasses in less productive soils of Southern European Union regions, results show that cultivation on low quality land is in general more costly per ton of produced biomass, in spite of the lower opportunity cost (rent) of land. However, the study also shows that production of energy crops in degraded land can increase and diversify farmers' income through access to the bio-based markets (i.e. biomaterial or bioenergy industries, amongst others), and the possibility to exploit less productive land with limited value for conventional agriculture. In the study, it was shown that production of domestic heat with pellets is a competing solution when compared with the conventional fossil fuel (oil heating). In addition, if a full plant operation is feasible for more than 6 months per year, Combined Heat and Power (CHP) applications are also suitable and profitable. Moreover, profits can be maximized if subsidies and grants are attributed to the production and conversion of biomass to energy, if an additional remuneration for CO_2 abatement or other benefits are credited, and, if the cost of contaminated land remediation is included (Boléo et al. 2013).

Regarding socio-economic effects, cultivation of energy crops in contaminated land is also beneficial. It contributes to diversify farming activities, providing new opportunities for farmers and to expand the rural economy (Boléo et al. 2013). In fact, the use of contaminated land for biomass production still involves much controversy, and not always have social acceptance, but simultaneously this approach contributes to the remediation of these soils, restoring its ecosystem function and services, and thus reducing human and environment exposure to pollutants. Also, the reduction of GHG emissions and thus the human exposure to their effects on health and to the environment represents an added benefit. Production of energy crops in contaminated land, besides providing an additional economic income, present also a positive gain in terms of employment in small and medium-size enterprises and on the employment in rural areas, especially if these areas are less productive due to contaminated soils. The positive influence on employment is also due to the contribution towards avoiding a rural exodus and to the contribution towards a more balanced rural development (Boléo et al. 2013).

The production of energy crops in heavy metals contaminated soils rely also on the immobilization of contaminants through sorption, precipitation, complexation or contaminant valence change (Ghosh and Singh 2005), preventing its loss to ground and surface waters. Several studies show that effective contaminants stabilization can be achieved by cultivation of energy crops in contaminated land. Barbosa et al. (2015a), in their work, showed that the response of giant reed and miscanthus in soils contaminated, limited the loss of heavy metals through percolated water, guaranteeing protection to the groundwater. Costa et al. (2013) evaluated the application to soils of Zn and Cu and its effect on giant reed growth, and verified that the radicular system of giant reed simultaneously absorbed and adsorbed over 90% of the pollutant load. The same pattern was also identified in *Miscanthus* spp. under the application of Zn, Cu, Ni and Cd (Bandarra et al. 2013; Lino et al. 2014) and kenaf in soils contaminated with Pb (Ho et al. 2008). Indian mustard, hybrid poplars and herbaceous plants also showed phytostabilization capacity to different heavy metals applied to fields (EPA 2000). In those works it was argued that heavy metals mobilized from the soil to the water were trapped in the root and rhizome system of those crops, reducing the diffuse pollution. Other bast and grass fiber crops (e.g. hemp, bamboo, soft rush and papyrus) could also be used successfully with the same purposes, as it was demonstrated in the review made by Barbosa et al. (2015b). Nevertheless, care should be taken regarding the emission of exudates from plants, namely organic acids, because they might also stimulate the migration of pollutants to water and groundwater due to mobilization (Fernando 2005). Increment of pH and soil organic matter (e.g. though application of biosolids) and addition of carbonates and phosphates are examples of actions, single or combined, that help to enhance the immobilization of toxic contaminants, such as metals (Fernando 2005). Yet, when phytostabilization is procured the contaminant still persists in the soil and therefore regular monitoring is required (Ghosh and Singh 2005).

Benefits and opportunities derived from the introduction of energy crops in a contaminated field are also associated with the increment in biological and landscape diversity. Although the yield loss and the reduced biomass density that can be observed in contaminated soils, by opposition to non-contaminated soils, may result in less coverage for fauna and less soil litter. However, compared to degraded/contaminated land where the spontaneous vegetation is sparse due to severe loss of soil fertility, energy crops fields cultivated on this type of soils may have a strong positive impact on biodiversity and landscape values (Boléo et al. 2013). Compared to the sparse spontaneous vegetation growing in contaminated land, energy crops provide more complex structure and heterogeneity with a positive influence on its cover value for wildlife and enriching landscape values. The inclusion of energy crops in contaminated soils has the advantage of converting the contaminated sites into a more aesthetically appealing landscape, a fact that could find easier public interest and support (McIntyre 2003). The invasive character of the crops (e.g. giant reed), penalizes the impact associated with the establishment of energy crops in contaminated soils, but native crops and flowering crops (e.g. cardoon), provide benefits towards biodiversity and landscape values (Fernando et al. 2018). The reduced soil tillage and use of agrochemicals, the highly above and belowground biomass, higher increments in soil organic matter content and litter deposition, favors perennials compared to annuals, towards biodiversity and landscape values (Fernando et al. 2010b). But, cultivation of energy crops in contaminated soils may represent also a threat to biodiversity, due to the monoculture system (Fernando et al. 2018).

4 Risks Associated with the Production of Energy Crops in Contaminated Land

The production of energy crops in contaminated soils depends on the biomass yields and quality and also on the ability of the crops to remediate the soils, restoring its properties.

Energy crops yields in contaminated soils are usually lower than in non-contaminated fields, as it was observed by Boléo et al. (2015) with *Miscanthus* x *giganteus* Greef et Deu in zinc contaminated soils and by Papazoglou and Fernando (2017) with sugarbeet (*Beta vulgaris* L.) in nickel contaminated soils. Lower yields are due to the contamination levels essayed that affected the biomass growth and productivity. Reduced yields forces the energy balance to decrease, once outputs will be lower, and the amount of fossil energy saved/conserved, is also lower. This was observed in the study made by Schmidt et al. (2015), where several perennial crops were screening for bioenergy. Life cycle analysis results showed that the lower the yields, the lower the fossil energy saved. However, when contamination do not affect the productivity of the biomass, as it was observed by Sidella et al. (2016) with giant reed in lead contaminated soils, the amount of fossil energy saved/conserved will not be affected. This happens when the level of contamination in the soil is not high enough to induce toxicity, or, when the existing contaminants are less mobile, thus limiting its accumulation/degradation by the plants, and the toxicity patterns derived.

Lower productivities also have an impact on the amount of GHG emissions and on the land use. Lower yields imply a lower abatement of GHG, due to lower carbon sequestration by the biomass, which may limit future applications of biomass produced in less productive land according to the European policies. This was observed in the study made by Fernando et al. (2016a), which compared the GHG emissions abatement due to the production and use of several perennial crops in degraded soils, and also in the study made by Boléo et al. (2013) that assessed the environmental and socio-economic impact of the *Miscanthus* production in zinc contaminated soils. Likewise, lower yields imply the need of higher area of land for the production of the same quantity of biomass. This may result in conflicts with pasture lands and lands of high ecological value, rich in biodiversity (Beringer et al. 2011). Moreover, landowner preferences should not be ignored and may represent a constraint, as the area of the less productive land that owners are willing to supply for bioenergy crops production is far less than the amount estimated based on remote sensing (Skevas et al. 2016). In fact, in the study reported by Skevas et al. (2016), made in Southern Lower Michigan, most of the surveyed landowners were not interested in renting land for bioenergy crops production and those who were interested offered relatively little land for bioenergy crops, even at rental rates three times current levels. Additionally, willing landowners would prefer to grow a significant portion of these crops on cropland rather than non-crop on the less productive soils.

Contaminated soils are usually very poor in organic matter and in nutrients, namely nitrogen, phosphorus, potassium, etc. derived from the loss in structure and holding capacity (Fernando 2005). Production of energy crops in these soils may rely on a higher supply of artificial fertilizers, to overcome the soil deficiencies, which in turn depend on the input of mineral resources (Biewinga and Van Der Bijl 1996). Hence, a higher exploitation of mineral resources (especially phosphate and potassium fertilizers) might be needed, increasing the impact towards exhaustion of mineral ores. Higher nitrogen applied to the soil can also increase the magnitude of some environmental impacts, which includes contribution to acidification, due to the volatilization of ammonia (NH_3) and N oxides (NO_x), to greenhouse effect and ozone depletion, due to denitrification to nitrous oxide (N_2O), and to ground and surface water eutrophication, due to ammonium (NH_4^+) and nitrate (NO_3^-) leaching and runoff (Fernando et al. 2010b, 2011).

The yield loss associated with cultivation in contaminated land and the need for higher inputs (fertilizers, fuel, water) may also hamper the economic viability of the crops (Fischer et al. 2010). Soldatos (2015) examined the economic aspects of bioenergy production from perennial crops in less productive soils of South Europe and found out that the production cost increased in the order 8–17% from standard agricultural land to land less productive. Results indicated that the disadvantage of the increased need for inputs and the lower productivity was not counterbalanced by the low economic rent of the poorer soils. In contrast, the yield loss and the need of higher land area to meet the feedstock supply increases the number of direct jobs (labor) required (Domac et al. 2005).

Nitrogen oxides are also emitted if the biomass is combusted. When nitrogen uptake by the biomass from contaminated soils is similar to the one uptake by the biomass from non-contaminated soils, NO_x emissions due to combustion of biomass will be similar, and no additional environmental impacts will be observed. This hypothesis was confirmed by Boléo et al. (2013) in the work that studied the environmental and socio-economic impact assessment of the miscanthus production in Zn contaminated soils. However, lower yields may induce concentration of elements, such as nitrogen or potassium, in the biomass, lowering the biomass quality for processing. Because of the soil toxicity, crops may not reach a mature status. In the case of perennial crops, this will limit the translocation process of N, P and K to the rhizomes before harvest, and may compromise the resprout of shoots the following spring, because the crops will not be able to produce sufficient metabolic reserves in their rhizome system (Fernando and Oliveira 2005). In the study of Barbosa et al. (2013), the phytoremediation response of giant reed in soils contaminated with zinc and chromium was evaluated. The contamination induced a yield loss and the harvested biomass presented a higher ash, nitrogen and potassium content, among other elements studied. Those characteristics may limit its industrial use: NO_x emissions from the combustion of biomass may increase; the ash produced has consequences on the handling and use/disposal of ash residues from biomass combustion plants; and fused agglomerates and slag deposits may increase, accelerating metal wastage of furnace and boilers components, reducing equipment's life (Livingston and Babcock 2006; Van der Weijde et al. 2017).

5 Concluding Remarks

The mitigation of potential risks and land scarcity due to land use change can be addressed by restoring functional and ecosystem services of contaminated land. The production of plants, namely energy crops, in heavy metals contaminated soils could be a sustainable, cost-effective option, with dual goals, pollution reduction and production of marketable biomass. The main advantage of the *in situ* approach, and in particular for soils contaminated, refers to the fact that it is not needed the removal of the soil, meaning that soils properties as fertility, structure, porosity and texture could be preserved and in some cases improved at the same time that contamination is remediated. It is consider also a "green" technology when compared with physico-chemical technologies because many of the impacts of the soil removal could be avoided. This technology could be applied to one or to a mixture of contaminants and because of that it is possible to apply in sites not ready to be remediated by other methods. Yet, the remediation success and the biomass being produced is also dependent on the concentration of the contaminants and the physical characteristics of the site.

According to several authors, energy crops have demonstrated the capacity to be tolerant to soil contamination, being able to uptake/degrade those pollutants, alleviating the soil contamination and restoring ecosystem services. At the same time, they represent a promising option for the bioeconomy, notably the biorefinery and bioenergy industries, providing additional revenue to farmers. Production of energy crops in contaminated land also avoids changes in arable land, helping to mitigate the food vs. biofuel debate. Other benefits rely on the fact that the knowledge applied to food crops as well in silvicultural practices can be applied to energy crops in contaminated soils. Moreover, the prospect of the valorization of the biomass could lessen the financial costs associated with the soil remediation.

However, sustainability of energy crops production in contaminated soils depends on its above and below productivity, on the ability of the crops to restore value to the land on the uptake and accumulation/degradation rates for the substance of interest and on the quality of the biomass being produced. An evaluation of several works presented in the literature suggest that the production of energy crops in contaminated soils have positive and less positive aspects over production in non-contaminated soils. The productivity loss in contaminated soils diminishes the energy, and the greenhouse savings but the presence of vegetation may contribute to improve the quality of soil and water and the biological and landscape diversity. Uptake of contaminants by biomass, if desirable to improve the phytotreatment, reducing the time needed for decontamination, by opposition, can be detrimental for its use and economic valorization. Another drawback of this technique is the time needed if the remediation of soils is to be accomplish, once several growing cycles are required to complete it, by comparison with the traditional physical-chemical techniques.

As part of active planning, bench-scale treatability studies should be conducted prior to field implementation. These studies represent a cost-effective tool for simultaneously evaluating multiple variables, optimizing performance and ultimately reducing environmental, social and economic costs. Moreover, growing energy crops in contaminated land should be integrated with landscape planning or regional land use development. This integration is particularly important when remediation by energy crops of the contaminated land is also a target as this is a prolonged process.

References

Arora K, Sharma S, Monti A (2016) Bio-remediation of Pb and Cd polluted soils by switchgrass: a case study in India. Int J Phytoremediation 18:704–709

Baird C (1999) Environmental chemistry, 2nd edn. WH Freeman and Company, New York

Balan V, Kumar S, Bals B, Chubdawat S, Jin M, Dale B (2012) Biochemical and Thermochemical conversion of switchgrass to biofuels. In: Monti A (ed) Switchgrass, a valuable biomass crop for energy. Green Energy and Technology, Springer-Verlag, London, pp 153–185

Bandarra V, Fernando AL, Boléo S, Barbosa B, Costa J, Sidella S, Duarte MP, Mendes B (2013) Growth, productivity and biomass quality of three Miscanthus genotypes irrigated with Zn and cu contaminated wastewaters. In: Eldrup A, Baxter D, Grassi A, Helm P (eds) Proceedings of the 21th European biomass conference and exhibition, setting the course for a biobased economy. ETA-Renewable Energies and WIP-Renewable Energies, Copenhagen, pp 147–150

Barbosa B, Boléo S, Sidella S, Costa J, Duarte MP, Mendes B, Cosentino SL, Fernando AL (2015a) Phytoremediation of heavy metal-contaminated soils using the perennial energy crops *Miscanthus* spp. and *Arundo donax* L. BioEnerg Res 8:1500–1511

Barbosa B, Costa J, Boléo S, Duarte MP, Fernando AL (2016) Phytoremediation of inorganic compounds. In: Ribeiro AB, Mateus EP, Couto N (eds) Electrokinetics across disciplines and continents - new strategies for sustainable development. Springer International Publishing, Switzerland, pp 373–400

Barbosa B, Costa J, Fernando AL, Papazoglou EG (2015b) Wastewater reuse for fiber crops cultivation as a strategy to mitigate desertification. Ind Crop Prod 68:17–12

Barbosa B, Fernando AL, Lino J, Costa J, Sidella S, Boléo S, Bandarra V, Duarte MP, Mendes B (2013) Phytoremediation response of *Arundo donax* L. in soils contaminated with zinc and chromium. In: Eldrup A, Baxter D, Grassi A, Helm P (eds) Proceedings of the 21st European biomass conference and exhibition, setting the course for a biobased economy. ETA-Renewable Energies and WIP-Renewable Energies, Copenhagen, pp 315–318

Benjamin M, Honeyman B (1992) Trace metals. In: Butcher S, Charlson R, Orians G, Wolfe G (eds) Global biogeochemical cycles. Academic Press Limited, San Diego

Beringer T, Lucht W, Schaphoff S (2011) Bioenergy production potential of global biomass plantations under environmental and agricultural constraints. GCB Bioenerg 3:299–312

Berti MT, Gesch RW, Eynck C, Anderson J, Cermak S (2016) Camelina uses, genetics, genomics, production and management. Ind Crop Prod 94:690–710

Biewinga EE, Van Der Bijl G (1996) Sustainability of energy crops in Europe: a methodology developed and applied. CLM, Utrecht

Bini C, Wahsha M, Fontana S, Maleci L (2012) Effects of heavy metals on morphological characteristics of Taraxacum officinale web growing on mine soils in NE Italy. J Geochem Explor 123:101–108

Bjelková M, Tejklová E, Griga M, Zajíková I, Genurová V (2001) Flax, linseed and hemp in phytoremediation, natural Fibres (Poznan) – special edition:In: Proceeding of the 2nd global workshop Bast plants in the new millennium. Borovets, Bulgaria, p 285

Boléo S, Fernando AL, Barbosa B, Costa J, Duarte MP, Mendes B (2015) Remediation of soils contaminated with zinc by *Miscanthus*. In: Vilarinho C, Castro F, Russo M (eds) WASTES 2015–solutions, treatments and opportunities: selected papers from the 3rd edition of the international conference on wastes: solution, treatments and opportunities. CRC Press, Taylor & Francis Group, Viana do Castelo, pp 37–42

Boléo S, Fernando AL, Duarte MP, Mendes B (2013) Environmental and socio-economic impact assessment of the *Miscanthus* production in Zn contaminated soils. In: Castro F, Vilarinho C, Carvalho J, Castro A, Araújo J, Pedro A (eds) Book of proceedings 2nd international conference: wastes: solutions, treatments and opportunities. CVR, Centro de Valorização de Resíduos, Braga, pp 657–662

Börjesson P (1999) Environmental effects of energy crop cultivation in Sweden—I: identification and quantification. Biomass Bioenergy 16:137–154

Bosco S, o Di Nasso NN, Roncucci N, Mazzoncini M, Bonari E (2016) Environmental performances of giant reed (*Arundo donax* L.) cultivated in fertile and marginal lands: a case study in the Mediterranean. Eur J Agron 78:20–31

Brandão M, Milà i Canals L, Clift R (2010) Soil organic carbon changes in the cultivation of energy crops: implications for GHG balances and soil quality for use in LCA. Biomass Bioenergy 35:2323–2336

Catroga A, Fernando A, Oliveira JS (2005) Effects on growth, productivity and biomass quality of Kenaf of soils contaminated with heavy metals. In: Sjunnesson L, Carrasco JE, Helm P, Grassi A (eds) Biomass for energy, industry and climate protection - proceedings of the 14th European Biomass Conference & Exhibition. ETA-Florence and WIP-Munich, Paris, pp 149–152

Chiaramonti D, Grimm H, El Bassam N, Cendagorta M (2000) Energy crops and bioenergy for rescuing deserting coastal area by desalination: feasibly study. Bioresour Technol 72:131–146

Clemente R, Walker DJ, Bernal MP (2005) Uptake of heavy metals and as by Brassica juncea grown in a contaminated soil in Aznalcóllar (Spain): the effect of soil amendments. Environ Pollut 138:46–58

Cortina J, Amat B, Castillo V, Fuentes D, Maestre F, Padilla F, Rojo L (2011) The restoration of vegetation cover in the semi-arid Iberian southeast. J Arid Environ 75:1377–1384

Cosentino SL, Copani V, Scalici G, Scordia D, Testa G (2015) Soil Erosion mitigation by perennial species under Mediterranean environment. Bioenergy Res 8:1538–1547

Costa C, Jesus-Rydin C (2001) Site investigation on heavy metals contaminated ground in Estarreja – Portugal. Eng Geol 60:39–47

Costa J, Fernando AL, Coutinho M, Barbosa B, Sidella S, Boléo S, Bandarra V, Duarte MP, Mendes B (2013) Growth, productivity and biomass quality of Arundo irrigated with Zn and Cu contaminated wastewaters. In: Eldrup A, Baxter D, Grassi A, Helm P (eds) Proceedings of the 21st European biomass conference and exhibition, setting the course for a biobased economy. ETA-Renewable Energies and WIP-Renewable Energies, Copenhagen, pp 308–310

Dauber J, Brown C, Fernando AL, Finnan J, Krasuska E, Ponitka J, Styles D, Thrän D, Van Groeningen KJ, Weih M, Zah R (2012) Bioenergy from "surplus" land: environmental and socio-economic implications. BioRisk 7:5–50

Domac J, Richards K, Risovic S (2005) Socio-economic drivers in implementing bioenergy projects. Biomass Bioenergy 28:97–106

Duggan J (2005) The potential for landfill leachate treatment using willows in the UK—A critical review. Resour Conserv Recy 45:97–113

EPA (2000) Introduction to phytoremediation, Cincinnati. EPA/600/R-99/107

Epelde L, Mijangos I, Becerril JM, Garbisu C (2009) Soil microbial community as bioindicator of the recovery of soil functioning derived from metal phytoextraction with sorghum. Soil Biol Biochem 41:1788–1794

European Commission (2009) Towards a better targeting of the aid to farmers in areas with natural handicaps, SEC (2009) 450. COM 2009:161

European Commission, Joint Research Centre (2013) Assessing the risk of farmland abandonment in the EU, Institute for Environment and Sustainability, report EUR 25783 EN. Luxembourg Publications, Office of the European Union

Evangelou MW, Robinson BH, Günthardt-Goerg MS, Schulin R (2013) Metal uptake and allocation in trees grown on contaminated land: implications for biomass production. Int J Phytoremediation 15:77–90

Fernando A, Oliveira JS (2004) Effects on growth, productivity and biomass quality of Miscanthus x giganteus of soils contaminated with heavy metals. In: Van Swaaij WPM, Fjällström T, Helm P, Grassi A (eds) Biomass for energy, industry and climate protection - proceedings of the 2nd world biomass conference, Rome ETA-Florence and WIP-Munich, p 387-390

Fernando AL (2013) Miscanthus for a sustainable development: how much carbon is captured in the soil? In: Eldrup A, Baxter D, Grassi A, Helm P (eds) Proceedings of the 21st European biomass conference and exhibition, setting the course for a biobased economy. ETA-Renewable Energies and WIP-Renewable Energies, Copenhagen, pp 1842–1843

Fernando AL, Barbosa B, Costa J, Alexopoulou E (2016a) Perennial grass production opportunities and constraints on marginal soils. In: Faaij APC, Baxter D, Grassi A, Helm P (eds) Proceedings of the 24rd European biomass conference and exhibition, setting the course for a biobased economy. ETA-Florence Renewable Energies, Amsterdam, pp 133–137

Fernando AL, Barbosa B, Costa J, Papazoglou EG (2016b) Giant reed (Arundo donax L.): a multipurpose crop bridging phytoremediation with sustainable bio-economy. In: Prasad MNV (ed) Bioremediation and bioeconomy. Elsevier Inc, UK, pp 77–95

Fernando AL, Boléo S, Barbosa B, Costa J, Duarte MP, Monti A (2016c) Assessment of site-specific environmental impacts of bioenergy and bio-based products from perennial grasses cultivated on marginal land in the Mediterranean region. In: APC F, Baxter D, Grassi A, Helm

P (eds) Proceedings of the 24rd European biomass conference and exhibition, setting the course for a biobased economy. ETA-Florence Renewable Energies, Amsterdam, pp 1525–1529

Fernando AL, Boléo S, Barbosa B, Costa J, Duarte MP, Monti A (2015) Perennial grass production opportunities on marginal Mediterranean land. Bioenergy Res 8:1523–1537

Fernando AL, Boléo S, Barbosa B, Costa J, Lino J, Tavares C, Sidella S, Duarte MP, Mendes B (2014) How sustainable is the production of energy crops in heavy metal contaminated soils? In: Hoffmann C, Baxter D, Maniatis K, Grassi A, Helm P (eds) Proceedings of the 22th European biomass conference and exhibition, setting the course for a biobased economy. ETA-Renewable Energies, Hamburg, pp 1593–1596

Fernando AL, Costa J, Barbosa B, Monti A, Rettenmaier N (2018) Environmental impact assessment of perennial crops cultivation on marginal soils in the Mediterranean region. Biomass Bioenergy 111:174–186. https://doi.org/10.1016/j.biombioe.2017.04.005

Fernando AL, Duarte MP, Almeida J, Boléo S, Di Virgilio N, Mendes B (2010a) The influence of crop management in the environmental impact of energy crops production. In: Spitzer J, Dallemand JF, Baxter D, Ossenbrink H, Grassi A, Helm P (eds) Proceedings of the 18th European biomass conference and exhibition, from research to industry and markets, Lyon ETA-Renewable Energies and WIP-Renewable Energies, p 2275-2279

Fernando AL, Duarte MP, Almeida J, Boléo S, Mendes B (2010b) Environmental impact assessment (EIA) of energy crops production in Europe. Biofuels Bioprod Biorefin 4:594–604

Fernando AL, Duarte MP, Almeida J, Boléo S, Mendes B (2011) Environmental pros and cons of energy crops cultivation in Europe. In: Faulstich M, Ossenbrink H, Dallemand JF, Baxter D, Grassi A, Helm P (eds) Proceedings of the 19th European biomass conference and exhibition, from research to industry and markets. ETA-Florence Renewable Energies, Berlin, pp 38–42

Fernando AL, Oliveira JFS (2005) Caracterização do potencial da planta Miscanthus x giganteus em Portugal para fins energéticos e industriais. Biologia Vegetal e Agro-Industrial 2:195–204

Fernando ALAC (2005) Fitorremediação por Miscanthus x giganteus de solos contaminados com metais pesados, PhD thesis. FCT/UNL, Lisboa

Fischer G, Prieler S, van Velthuizen H, Berndes G, Faaij A, Londo M, de Wit M (2010) Biofuel production potentials in Europe: sustainable use of cultivated land and pastures, part II: land use scenarios. Biomass Bioenergy 34:173–187

Ghosh M, Singh S (2005) A review on phytoremediation of heavy metals and utilization of its byproducts. Appl Ecol Environ Res 3:1–18

Giachetti G, Sebastiani L (2006) Metal accumulation in poplar plant grown with industrial wastes. Chemosphere 64:446–454

Ginneken LV, Meers E, Guisson R, Ruttens A, Elst K, Tack FMG, Vangronsveld J, Diels L, Dejonghe W (2007) Phytoremediation for heavy metal-contaminated soils combined with bioenergy production. J Environ Eng Landsc 15:227–236

Giuseppe DD, Antisari LV, Ferronato C, Branchini G (2014) New insights on mobility and bioavailability of heavy metals in soils of the Padanian alluvial plain (Ferrara Province, northern Italy). Chem Erde – Geochem 74:615–623

Grabowska L, Baraniecki P (1997) Three year results on utilization soil polluted by copper-producing industry. In: Proceedings of the flax and other bast plants Symp. Natural Fibres Spec. Ed. INF, Poznan, pp 123–131

Guo ZH, Miao XF (2010) Growth changes and tissues anatomical characteristics of giant reed (*Arundo donax* L.) in soil contaminated with arsenic, cadmium and lead. J Cent S Univ Technol 17:770–777

Hammer D, Kayser A, Keller C (2003) Phytoextraction of Cd and Zn with *Salix viminalis* in field trials. Soil Use Manage 3:187–192

Herpin U, Berlekamp J, Markert B, Wolterbeek B, Grodzinska K, Siewers U, Lieth H, Weckert V (1996) The distribution of heavy metals in a transect of the three states the Netherlands, Germany and Poland, determined with the aid of moss monitoring. Sci Total Environ 187:185–198

Ho W, Ang L, Lee D (2008) Assessment of Pb uptake, translocation and immobilization in kenaf (*Hibiscus cannabinus* L.) for phytoremediation of sand tailings. J Environ Sci 20:1341–1347

Hüffmeyer N, Klasmeier J, Matthies M (2009) Geo-referenced modeling of zinc concentrations in the Ruhr river basin (Germany) using the model GREAT-ER. Sci Total Environ 407:2296–2305

Kabata-Pendias A (2011) Trace elements in soils and plants, 4th edn. CRC Press, INc., Boca Raton

Leung HM, Ye ZH, Wong MH (2007) Survival strategies of plants associated with arbuscular mycorrhizal fungi on toxic mine tailings. Chemosphere 66:905–915

Lewandowski I, Schmidt U, Londo M, Faaij A (2006) The economic value of the phytoremediation function – assessed by the example of cadmium remediation by willow (*Salix* ssp). Agric Syst 89:68–89

Linger P, Müssig J, Fischer H, Kobert J (2002) Industrial hemp (*Cannabis sativa* L.) growing on heavy metal contaminated soil: fibre quality and phytoremediation potential. Ind Crop Prod 16:33–42

Lino J, Fernando AL, Barbosa B, Boléo S, Costa J, Duarte MP, Mendes B (2014) Phytoremediation of Cd and Ni contaminated wastewaters by Miscanthus. In: Hoffmann C, Baxter D, Maniatis K, Grassi A, Helm P (eds) Proceedings of the 22th European biomass conference and exhibition, setting the course for a biobased economy. ETA-Renewable Energies, Hamburg, pp 303–307

Lips SJJ, Van Dam JEG (2013) Kenaf fibre crop for bioeconomic industrial development. In: Monti A and Alexopoulou E (ed) Kenaf: a multi-purpose crop for several industrial applications, green energy and technology, Springer-Verlag, London, p 105-143

Livingston B, Babcock M (2006) Ash related issues in biomass combustion. ThermalNet workshop, Glasgow

McIntyre T (2003) Phytoremediation of heavy metals from soils. Adv Biochem Eng Biotechnol 78:97–123

Meers E, Ruttens A, Hopgood M, Lesage E, Tack F (2005) Potential of Brassica rapa, Cannabis sativa, Helianthus annuus and Zea mays for phytoextraction of heavy metals from calcareous dredged sediment derived soils. Chemosphere 61:561–572

Meers E, Vandecasteele B, Ruttens A, Vangronsveld J, Tack F (2007) Potential of five willow species (*Salix* spp.) for phytoremediation of heavy metals. Environ Exp Bot 60:57–68

Mirza N, Pervez A, Mahmood Q, Shah M, Shafqat M (2011) Ecological restoration of arsenic contaminated soil by *Arundo donax* L. Ecol Eng 37:1949–1956

Monti A, Zegada-Lizarazu W (2016) Sixteen-year biomass yield and soil carbon storage of giant reed (*Arundo donax* L.) grown under variable nitrogen fertilization rates. Bioenergy Res 9:248–256

Nicoletti G, Arcuri N, Nicoletti G, Bruno R (2015) A technical and environmental comparison between hydrogen and some fossil fuels. Energy Convers Manag 89:205–213

Niu Z, Sun L, Sun T, Li Y, Wang H (2007) Evaluation of phytoextracting cadmium and lead by sunflower, ricinus, alfalfa and mustard in hydroponic culture. J Environ Sci 19:961–967

Nsanganwimana F, Marchand L, Douay F, Mench M (2014a) *Arundo donax* L., a candidate for phytomanaging water and soils contaminated by trace elements and producing plant-based feedstock. A review. Int J Phytoremediation 16:982–1017

Nsanganwimana F, Pourrut B, Mench M, Douay F (2014b) Suitability of Miscanthus species for managing inorganic and organic contaminated land and restoring ecosystem services. A review. J Environ Manag 143:123–134

Pandey VP, Bajpai O, Singh N (2016) Energy crops in sustainable phytoremediation. Renew Sust Energ Rev 54:58–73

Papazoglou E, Karantounias G, Vemmos S, Bouranis D (2005) Photosynthesis and growth responses of giant reed (*Arundo donax* L.) to the heavy metals Cd and Ni. Environ Int 31:243–249

Papazoglou EG, Fernando AL (2017) Preliminary studies on the growth, tolerance and phytoremediation ability of sugarbeet (*Beta vulgaris* L.) grown on heavy metal contaminated soil. Ind Crop Prod 107:463–471. https://doi.org/10.1016/j.indcrop.2017.06.051

Pidlisnyuk V, Stefanovska T, Lewis EE, Erickson LE, Davis LC (2014) Miscanthus as a productive biofuel crop for Phytoremediation. Crit Rev Plant Sci 33:1–19

Prasad M (2004) Heavy metal stress in plants, from biomolecules to ecosystems, 2nd edn. Springer, Hyderabad

Pratas J, Favas PJC, D'Souza R, Varun M, Paul MS (2013) Phytoremedial assessment of flora tolerant to heavy metals in the contaminated soils of an abandoned Pb mine in Central Portugal. Chemosphere 90:2216–2225

Rajkumar M, Freitas H (2008) Influence of metal resistant-plant growth-promoting bacteria on the growth of *Ricinus communis* in soil contaminated with heavy metals. Chemosphere 71:834–842

Redovniković IR, De Marco A, Proietti C, Hanousek K, Sedak M, Bilandžić N, Jakovljević T (2017) Poplar response to cadmium and lead soil contamination. Ecotoxicol Environ Saf 144:482–489

Russell DJ, Alberti G (1998) Effects of long-term, geogenic heavy metal contamination on soil organic matter and microarthropod communities, in particular Collembola. Appl Soil Ecol 9:483–488

Sabra N, Dubourguier H, Hamieh T (2011) Sequential extraction and particle size analysis of heavy metals in sediments dredged from the Deûle Canal, France. The Open Environ Eng J 4:11–17

Schmidt T, Fernando AL, Monti A, Rettenmaier N (2015) Life cycle assessment of bioenergy and bio-based products from perennial grasses cultivated on marginal land in the Mediterranean region. BioEnerg Res 8:1548–1561

Schröder P, Herzig R, Bojinov B, Ruttens A, Nehnevajova E, Stamatiadis S, Memon A, Vassilev A, Caviezel M, Vangronsveld J (2008) Bioenergy to save the world. Producing novel energy plants for growth on abandoned land. Environ Sci Pollut Res Int 15:196–204

Shcheglov AI, Olga B, Tsvetnova OB, Klyashtorin A (2014) The fate of Cs-137 in forest soils of Russian Federation and Ukraine contaminated due to the Chernobyl accident. J Geochem Explor 142:75–81

Sheng XF, Xia JJ, Jiang CY, He LY, Qian M (2008) Characterization of heavy metal-resistant endophytic bacteria from rape (*Brassica napus*) roots and their potential in promoting the growth and lead accumulation of rape. Environ Pollut 156:1164–1170

Sidella S, Barbosa B, Costa J, Cosentino SL, Fernando AL (2016) Screening of Giant reed clones for Phytoremediation of lead contaminated soils. In: Barth S, Murphy-Bokern D, Kalinina O, Taylor G, Jones M (eds) Perennial biomass crops for a resource constrained world. Springer International Publishing, Switzerland, pp 191–197

Sidella S, Fernando AL, Barbosa B, Costa J, Boléo S, Bandarra V, Duarte MP, Mendes B, Cosentino SL (2013) Phytoremediation response of *Arundo donax* in soils contaminated with lead. In: Eldrup A, Baxter D, Grassi A, Helm P (eds) Proceedings of the 21st European biomass conference and exhibition, setting the course for a biobased economy. ETA-Renewable Energies and WIP-Renewable Energies, Copenhagen, pp 385–387

Skevas T, Hayden NJ, Swinton SM, Lupi F (2016) Landowner willingness to supply marginal land for bioenergy production. Land Use Policy 50:507–517

Soldatos P (2015) Economic aspects of bioenergy production from perennial grasses in marginal lands of South Europe. Bioenergy Res 8:1562–1573

Stewart CE, Follett RF, Pruessner EG, Varvel GE, Vogel KP, Mitchell RB (2015) Nitrogen and harvest effects on soil properties under rainfed switchgrass and no-till corn over 9 years: implications for soil quality. GCB Bioenerg 7:288–301

UWE (2013) Science for environment policy in-depth report: soil contamination: impacts on human health. Science Communication Unit, University of the West of England, Bristol Report produced for the European Commission DG Environment. September 2013

Van de Weijde T, Kiesel A, Iqbal Y, Muylle H, Dolstra O, Visser RGF, Lewandowski I, Trindade LM (2017) Evaluation of Miscanthus sinensis biomass quality as feedstock for conversion into different bioenergy products. GCB Bioenerg 9:176–190

Vystavna Y, Rushenko L, Diadin D, Klymenko O, Klymenko M (2014) Trace metals in wine and vineyard environment in southern Ukraine. Food Chem 146:339–344

Wang Z, Zhang Y, Huang Z, Huang L (2008) Antioxidative response of metal-accumulator and non-accumulator plants under cadmium stress. Plant Soil 310:137–149

Wuana RA, Okieimen FE (2011) Heavy metals in contaminated soils: a review of sources, chemistry, risks and best available strategies for remediation. ISRN Ecology. https://doi.org/10.5402/2011/402647

Yang X, Feng Y, He Z, Stoffella P (2005) Molecular mechanisms of heavy metal hyperacumulation and phytoremediation. J Trace Elem Med Bio 18:339–353

Yordanova I, Staneva D, Misheva L, Bineva T, Banov M (2014) Technogenic radionuclides in undisturbed Bulgarian soils. J Geochem Explor 142:69–74

Zárubová P, Hejcman M, Vondráčková S, Mrnka L, Száková J, Tlustoš P (2015) Distribution of P, K, Ca, Mg, Cd, Cu, Fe, Mn, Pb and Zn in wood and bark age classes of willows and poplars used for phytoextraction on soils contaminated by risk elements. Environ Sci Pollut Res Int 22:18801–18813

Zatta A, Clifton-Brown J, Robson P, Hastings A, Monti A (2014) Land use change from C3 grassland to C4 Miscanthus: effects on soil carbon content and estimated mitigation benefit after six years. GCB Bioenergy 6:360–370

Zegada-Lizarazu W, Monti A (2011) Energy crops in rotation. A review. Biomass Bioenergy 35:12–25

Farmers' Acreage Responses to the Expansion of the Sugarcane Ethanol Industry: The Case of Goiás and Mato Grosso Do Sul, Brazil

Gabriel Granco, Marcellus Caldas, Allen Featherstone, Ana Cláudia Sant'Anna, and Jason Bergtold

Abstract From 2005 to 2012 sugarcane planted area increased by 54% in Brazil, reaching 9 million ha. This expansion was stronger in the Brazilian Cerrado, especially in the states of Goiás and Mato Grosso do Sul which are the new frontier of sugarcane production. The rapid expansion of sugarcane production in Brazil has the potential to reorganize the agricultural production landscape. Previous studies that examined the expansion trend and production system at a larger scale found evidence for the transition to a sugarcane producing region. However, little is known on how farmers decide which agricultural production to pursue and which land use to replace in the new frontier of sugarcane production. The goal of this chapter is to analyze farmers' acreage response during the proliferation of the sugarcane industry into the new production frontier. More specifically, we estimate a partial adjustment model to examine farmers' decisions toward sugarcane production in the states of Goiás and Mato Grosso do Sul. We estimate acreage response at county level using a partial adjustment framework. Estimates found that price of cattle has the largest cross-price elasticity with sugarcane acreage. In addition, the results suggest that acreage of sugarcane and soybean double-crop are positively correlated.

G. Granco (✉)
Stroud Water Research Center, Avondale, PA, USA
e-mail: ggranco@ksu.edu

M. Caldas
Kansas State University, Department of Geography, Manhattan, KS, USA

A. Featherstone · J. Bergtold
Kansas State University, Department of Agricultural Economics, Manhattan, KS, USA

A. C. Sant'Anna
The Ohio State University, Department of Agricultural, Environmental, and Development Economics, Columbus, OH, USA

© Springer International Publishing AG, part of Springer Nature 2018
R. Li, A. Monti (eds.), *Land Allocation for Biomass Crops*,
https://doi.org/10.1007/978-3-319-74536-7_6

1 Introduction

Brazil is the focus of worldwide attention because of its growing agricultural production and large reserves of native vegetation and diverse biomes. The biggest concern is the conversion of native vegetation to agricultural land. Since the year 2000, the agricultural land in Brazil increased by 21 million ha ((Food and Agricultural Organization (FAO), FAOSTAT: Land use, FAO Statistics Division, available at http://faostat.fao.org). Currently, the attention is focused on the Amazon biome. However, another biome in Brazil has endured more conversion of its natural cover than the Amazon without receiving as much attention (Strassburg et al. 2017). This is the case of the Cerrado biome (Brazilian Savanna), the second largest biome in Brazil after the Amazon and regarded as the most biodiverse savanna in the world (Myers et al. 2000).

The Cerrado natural vegetation originally extended over more than 203 million ha, however, by 2013, only 54% remains under its natural cover (Brazil 2015). The main anthropogenic modifications to this biome are the conversion of native vegetation to cropland and to pastureland which are threatening the biome's biodiversity and ecosystem services. The threats to the Cerrado lead to the classification of this biome as a global biodiversity hotspot (Carvalho et al. 2009; Myers et al. 2000).

The agricultural use of the Cerrado gained traction in the 1970s with the adaptation of soybeans to the Cerrado environment creating a new agricultural frontier. Later, this region became the breadbasket for Brazil (Barretto et al. 2013; Klink and Machado 2005; Rada 2013). This transformation is based on a capital-intensive approach to agricultural production (Brannstrom et al. 2008; Ferreira et al. 2016; Jepson 2006; Jepson et al. 2010; Silva and Miziara 2011). The traditional commercial agricultural land uses in the Cerrado are pasture to sustain cattle ranching and grain production, including soybeans and corn (Ferreira et al. 2013, 2016; Rodrigues and Miziara 2008; Sano et al. 2010). The Cerrado continues to be the agricultural production frontier in Brazil.

Since the 2000s, the Cerrado has seen the expansion of another frontier, the sugarcane ethanol frontier (Silva and Miziara 2011). The recent rise in demand for sugar and ethanol has stimulated the expansion of the sugarcane industry in Brazil, especially with the installation of new mills in the Cerrado (Granco et al. 2018). The movement of new mills in Goiás and Mato Grosso do Sul is a combination of several factors. First, the mills demand large and flat expanses of land – necessary to reduce the sugarcane procurement cost (Sant'Anna et al. 2016b), furthermore, these areas were available at low cost in these states (Granco et al. 2017). Second, the enforcement of environmental legislation in the state of São Paulo (Aguiar et al. 2011; IEA- Instituto de Economia Agrícola 2014) that hindered the expansion in that state. Third, the political and fiscal support from the governments in Goiás and Mato Grosso do Sul (Granco et al. 2017; Sant'Anna et al. 2016b). During the period between 2005 and 2013 (Fig. 1), the area planted to sugarcane increased by 430%, reaching 1.4 million ha while the number of mills reached 60 being 37 new units. Because of this expansion, these states represented 22% of sugarcane ethanol in Brazil in 2016 (CONAB 2016).

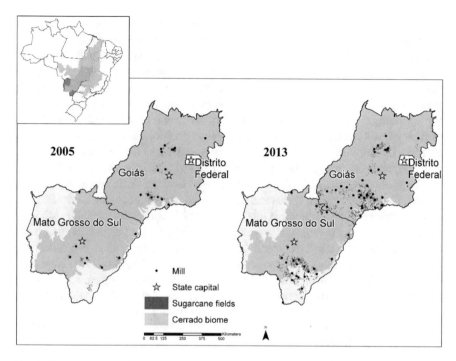

Fig. 1 Sugarcane expansion into Goiás and Mato Grosso Sul (Granco 2017)

This rapid expansion of sugarcane production in Brazil has the potential to reshape the agricultural production landscape (Goldemberg et al. 2014; Strassburg et al. 2014). The addition of a new cash-crop to the Cerrado biome has raised concerns over the sustainability of sugarcane and ethanol production (Bergtold et al. 2017; Walter et al. 2014). As a response to these concerns, the Brazilian government developed a land use policy to coordinate the sugarcane expansion, the Sugarcane Agroecological Zoning (SAZ). The SAZ defines areas that are suitable or non-suitable for sugarcane expansion considering natural conditions, mechanization, and previous land cover (Manzatto et al. 2009). The SAZ rules out any direct conversion of natural vegetation to create sugarcane fields. Moreover, the SAZ aims to reduce the impact of sugarcane expansion on food production by indicating a preference toward sugarcane in replacing degraded pastureland instead of cropland (Manzatto et al. 2009). This policy backing sugarcane substitution of degraded pastureland is supported by research results indicating that this land transition emits small amount of greenhouses gasses (GHG) and has a minor impact on food production (Alkimim et al. 2015; Cohn et al. 2014; Goldemberg et al. 2014; Leal et al. 2013; Strassburg et al. 2014). The SAZ indicates that Goiás and Mato Grosso do Sul have 35% of the areas suitable for sugarcane production accounting to more than 23 million hectares (Granco et al. 2017; Manzatto et al. 2009).

However, these studies do not analyze the farmers' land use decision-making process behind sugarcane expansion—more specifically, the factors affecting farmers' decisions to substitute pasture for sugarcane. A farmers' land use decision-making process can be considered as an optimization problem where farmers allocate land to the use that results in the highest economic return (Hennessy 2006). To maximize profits, farmers decide which agricultural production to pursue given the technological package available. Notably, farmers do not control all the factors in their decision-making process because those decisions are also conditioned by external factors such as government policies and laws, demand for agricultural products, climate, among others (Bergtold et al. 2014; Caldas et al. 2014, 2015).

The goal of this chapter is to examine farmers' land use decisions in the Cerrado, specifically in the states of Goiás and Mato Grosso do Sul, focusing on the increase of sugarcane area. To accomplish this goal, a partial-adjustment model is estimated for Cerrado acreage response at a county level using a panel regression that treats each land use decision as a result of profit maximization, thus estimating the optimal allocation of land (Carpentier and Letort 2014; Hausman 2012; Hendricks et al. 2014; Wu and Brorsen 1995). In this context, explanatory variables are the previous acreage for the main commercial land use and for the natural Cerrado vegetation, commodity prices and yields, and distance to the nearest mill. As a result, the model estimates the impact on land allocation and is conditional on changes in these factors.

2 Methods

2.1 Study Area

The Brazilian Cerrado encompasses 204 million ha, involving 11 states and the Federal District, including the states of Goiás and Mato Grosso do Sul. Originally, the Cerrado covered 98% of the Goiás territory (34 million ha) and more than 60% of the Mato Grosso do Sul territory (21 million ha) (Ministério do Meio Ambiente 2014). In these states, the Cerrado exhibits a variety of vegetation patterns, ranging from open grassland to closed woodland in a soil that is deep, well drained, and resistant to compaction—although it is acidic, with poor nutrient content, and a high concentration of aluminum (Brannstrom et al. 2008; Klink and Machado 2005).

Nevertheless, advances in agronomic technology allowed farmers to overcome soil deficiencies and transform the Cerrado into a breadbasket (Jepson et al. 2010). The main agricultural productions in this region are cattle, soybeans, and corn (Fig. 2). Considering the states of Goiás and Mato Grosso do Sul, pastureland is the main land use covering more than 26 million ha in 2013 (Brazil 2015), followed by soybeans with close to 5 million ha, while sugarcane covers around 1.4 million ha (CONAB 2016). Together, both states are on the forefront of sugarcane expansion in the Cerrado with 60 processing mills (36 in Goiás and 24 in Mato Grosso do Sul) in 33 and 21 counties, respectively.

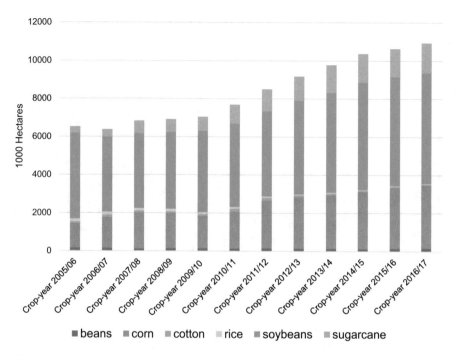

Fig. 2 Historical evolution of cropland by main crop products for the states of Goiás and Mato Grosso do Sul, for the crop-years of 2005/06 to 2016/17 (CONAB 2016)

The definition of the study area encompasses only the counties that are completely within the Cerrado biome in the states of Goiás and Mato Grosso do Sul. To identify these counties, first, the county borders are overlaid with the official shapefile delimiting the Cerrado; second, the counties are intersected with the PROBIO maps. The PROBIO program (*Projeto de Conservação e Utilização Sustentável da Diversidade Biológica Brasileira Conservation and Sustainable - Use of Brazilian Biological Diversity Project*) is a federal government effort to map and classify the remaining natural vegetation in Brazil as in 2002 (Ministério do Meio Ambiente 2008). The study area encompasses 243 counties, 219 in the state of Goiás and 24 in the state of Mato Grosso do Sul (Fig. 3).

This process is necessary for the next procedure that identifies the land use for these counties. The classification of land cover and land use in the Cerrado is not exempt from controversy (Brannstrom and Filippi 2008; Brazil 2015; Brown et al. 2013; Müller et al. 2015; Sano et al. 2008). One point of discussion is the classification of grassland into planted pastureland or native pastureland. This classification is fundamental for the study of land allocation and conservation of the Cerrado because pastureland is the main anthropogenic use in this biome (Brazil 2015).

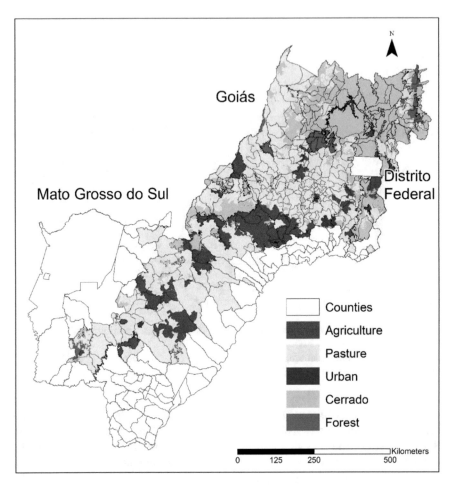

Fig. 3 Counties situated in the Cerrado biome and previously mapped by the PROBIO program

2.2 *Identifying Cerrado in a Time-Series of Land Use Maps*

Researchers have used remote sensing imagery to classify land cover–land use change in the Cerrado (Brannstrom and Filippi 2008; Brazil 2015; Brown et al. 2013; Müller et al. 2015; Sano et al. 2008). Given the dimension of this biome, most of the studies do not classify its totality; rather, they classify specific regions (spatial limitation). In addition, these studies focused on identifying land use change between two periods, thus only classifying land use for the start and end year of the study period. Developments in MODIS NDVI and EVI data sets now facilitate the detection of annual land use change (Brown et al. 2013). However, this advancement often groups different crops into a single category. Even though these studies advanced our knowledge on land use transition in the Cerrado given the spatial and temporal limitations, this approach does not support a robust economic model of farmers' land allocation decisions (Irwin and Geoghegan 2001).

To advance our understanding of the land allocation in the Cerrado, a time-series database of land cover–land use maps was constructed for the study area separating crops, pasture, and Cerrado for the period between 2005 and 2013. The creation of this database involved two processes. The first process was the land use–land cover (LULC) classification of MODIS NDVI 16-day imagery (Kastens et al. 2017). The classification was calibrated with field data obtained from interviews with farmers and landowners about their land use decisions following Wardlow et al. (2007), Brown et al. (2007), and Caldas et al. (2017). Each interviewee described the agricultural use of a specific field for the period of analysis generating ground-truth points. From the interviews, we gathered 1370 points of annual land cover history data representing 137 unique field sites and 10 years (2005–2014), of which 464 were pre-sugarcane fields. With this information, the following land use classes were classified: annual single crop, annual double crop, and pasture/cerrado.

The original imagery—consisting of the 250-m, 16-day composite MOD13Q1 Normalized Difference Vegetation Index (NDVI) data from the Moderate Imaging Spectroradiometer (MODIS) covering the study area for crop years 2005–2014—was downloaded from the Land Processes Distributed Active Archive Center (LP DAAC; https://lpdaac.usgs.gov/data_access). These data were reprojected to the WGS84 projection with a grid size of approximately 240 m. Annual MODIS NDVI profiles were extracted in correspondence with the ground reference data applying the pure pixel approach (Brown et al. 2007; Kastens et al. 2017; Wardlow et al. 2006; Wardlow and Egbert 2008). Using the 23-date MODIS profiles as independent variables and ground reference cover class as the dependent variable, a random forest (RF) classification model (Breiman 2001; Clark et al. 2010) was developed using the 'treebagger' function in MATLAB® (Kastens et al. 2017). One thousand trees were included in the forest, with each developed using a 5-element random subset of the 23 candidate predictors (MODIS time periods).

To estimate the expected error of the full RF model, 10 iterations of a one-year-holdout cross-validation (CV) exercise were used. For each iteration, 10 RF models were independently developed using unique 9-year subsets of the 10-year ground reference data set followed by an application of each model to ground reference data from its respective holdout year (Kastens et al. 2017). Aggregating the results across all 10 holdout years for each CV iteration, overall classification accuracies across the 10 iterations for the 3-class model outputs ranged from 80.8 to 81.9%, which increased to 90.3–91.4% when grouping the single-crop and double-crop classes.

Other classes were later added to the LULC maps to better represent the land uses options faced by farmers. The classes added are forest, sugarcane, urban, and water. The forest class was added using Global Forest Change data (2000–2014) (Hansen et al. 2013). The sugarcane class used a sugarcane classification map developed by the Canasat Program of the Brazilian National Institute for Space Research (INPE) (Rudorff et al. 2010), and the data were provided by the stewards of such data set. The urban and water class are static layers gathered from the Brazilian Institute of Geography and Statistics (IBGE) and burned into the maps.

The second process focused on splitting the class Cerrado/Pasture into individual classes: one for Cerrado and one for Pasture. The approach was to identify the areas that were classified as Cerrado in 2002 using PROBIO's classification (Sano et al. 2010) and to update these areas given annual deforestation maps for the period 2003–2014 by the Cerrado Warning Deforestation System (SIAD) (Ferreira et al. 2013; Ferreira et al. 2007). Thus, a Cerrado-only data set was created for 2003 to 2014. These data were integrated into the time-series database by splitting the Cerrado/Pasture category as follows: areas that were overlaid with the Cerrado-only data set were reclassified as Cerrado; areas outside of the Cerrado-only areas were reclassified as Pasture. Thus, the time-series database now supports the implementation of more advanced econometric analysis of land allocation by presenting annual change among the main commercial uses and native vegetation (Fig. 4).

Eight land classes are available: 1 – Sugarcane, 2 – Soybean single crop, 3 – Soybean double crop, 4 – Pasture, 5 – Cerrado, 6 – Planted forest, 7 – Urban, and 8 – Water. The first five classes represent the main land uses available in the study area. Sugarcane represents the new cash-crop that is expanding throughout the region (Bergtold et al. 2017; Granco et al. 2017; Shikida 2013). Soybean single crop is the traditional grain production that revolutionized the Cerrado in the 1970s (Jepson 2006; Rada 2013). Soybean double crop is the new production system that intensifies production with the intensification of inputs. The usual crop rotation is a short-season soybean, followed by corn (called *milho safrinha* by Brazilian farmers) (Arvor et al. 2012; Brown et al. 2013). Pasture is the main agricultural activity in the region with cattle ranching (Newberry 2014; Walker et al. 2009). Cattle ranching is an extensive practice with pastureland in the region exhibiting low cattle stocking per hectare and some level of degradation (Bustamante et al. 2012; Cohn et al. 2014; Martha et al. 2012; Strassburg et al. 2014). While the Cerrado class encompasses the native vegetation. The observed transitions from class to class can be used to identify the farmers' revealed preference for one land use over another.

2.3 Acreage Response Model

The Cerrado is still considered an agricultural frontier with areas to be opened, and farmers are working to decide the best use for the land. Farmers face the decision to maintain (or not) the native Cerrado on their farm (Bergtold et al. 2017). The farmers' land allocation decisions are considered a function of profit maximization bahavior, where profit is dependent on prices, production cost, distance to market and other transportation cost, terrain, among other factors (Peterson et al. 2014). The implementation of this model requires an annual identification of the main agricultural land uses (including natural vegetation) to characterize the land use transition.

Partial-adjustment models are an important tool for the study of farmers' acreage response to the changes in crop and input prices (Haile et al. 2016; Hendricks et al.

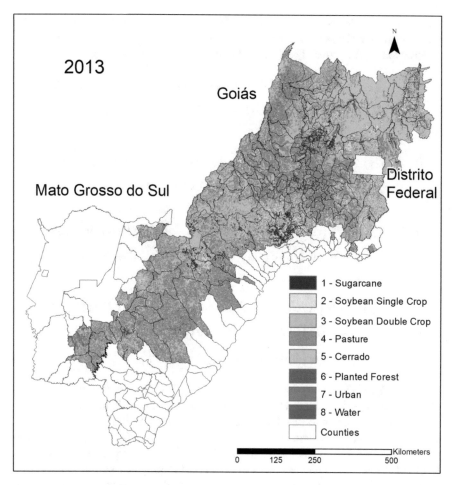

Fig. 4 Land use and land cover classification with specific classes for Cerrado and pasture as in 2013

2014). The foundational study in this field is by Nerlove (1956) which proposed that farmers' acreage response took longer to be completed than the short-run, resulting in more elastic estimates of acreage response. This claim was supported by demonstrating that farmers' price expectations were not constrained to the previous year's prices. Following Nerlove (1956), the partial-adjustment model has been applied for a large number of crops in several different countries or larger regions (Haile et al. 2016; Saddiq et al. 2014). Recently, partial adjustment models have been developed for examining biofuel crop expansion such as sugarcane and soybean acreage in Brazil (Hausman 2012) and soybean and corn acreage in the U.S. (Hendricks et al. 2014; Secchi et al. 2011).

The partial-adjustment model estimates current acreage as a function of last year's acreage, yield, and prices for the crop in question and for its land use substitutes. The inclusion of price and yield risk was not feasible given the available data. Sugarcane is a new cash-crop in the region with no history of price and yield specific to the study area. Thus, the following specifications are estimated:

$$
\begin{aligned}
\ln\left(ASC_{i,t}\right) = {} & \alpha \ln\left(ASC_{i,t-1}\right) + \beta_{1,l} \ln\left(A_{l,t-1}\right) + \beta_2 \ln\left(YLDSC_{i,t-1}\right) \\
& + \beta_{3,l} \ln\left(YLD_{l,t-1}\right) + \beta_4 \ln\left(DMill_{i,t-1}\right) + \nu_{i,j} + \mu_t + \varepsilon_{i,j,t}
\end{aligned}
\tag{1}
$$

$$
\begin{aligned}
\ln\left(ASC_{i,t}\right) = {} & \alpha \ln\left(ASC_{i,t-1}\right) + \beta_{1,l} \ln\left(A_{l,t-1}\right) + \beta_2 \ln\left(PSC_{i,t-1}\right) \\
& + \beta_{3,l} \ln\left(P_{l,t-1}\right) + \beta_4 \ln\left(DMill_{i,t-1}\right) + \nu_{i,j} + \mu_t + \varepsilon_{i,j,t}
\end{aligned}
\tag{2}
$$

$$
\begin{aligned}
\ln\left(ASC_{i,t}\right) = {} & \alpha \ln\left(ASC_{i,t-1}\right) + \beta_{1,l} \ln\left(A_{l,t-1}\right) + \beta_2 \ln\left(REVSC_{i,t-1}\right) \\
& + \beta_{3,l} \ln\left(REV_{l,t-1}\right) + \beta_4 \ln\left(DMill_{i,t-1}\right) + \nu_{i,j} + \mu_t + \varepsilon_{i,j,t}
\end{aligned}
\tag{3}
$$

$$
\begin{aligned}
\ln\left(ASC_{i,t}\right) = {} & \alpha \ln\left(ASC_{i,t-1}\right) + \beta_{1,l} \ln\left(A_{l,t-1}\right) + \beta_2 \ln\left(YLDSC_{i,t-1}\right) \\
& + \beta_{3,l} \ln\left(YLD_{l,t-1}\right) + \beta_4 \ln\left(PSC_{i,t-1}\right) + \beta_{5,l} \ln\left(P_{l,t-1}\right) \\
& + \beta_6 \ln\left(DMill_{i,t-1}\right) + \nu_{i,j} + \mu_t + \varepsilon_{i,j,t}
\end{aligned}
\tag{4}
$$

where $ASC_{i,t}$ is the county-level sugarcane-planted acreage in hectares; $ASC_{i,t-1}$ is the previous period county-level sugarcane-planted acreage in hectares; $A_{l,t-1}$ is the previous period county-level acreage vector for the l substitute uses (pastureland, grain–single crop, grain–double crop, and Cerrado); $YLDSC_{i,t-1}$ is the previous period county-level sugarcane yield; $YLD_{l,t-1}$ is the previous period county-level substitute uses yield vector; $PSC_{i,t-1}$ is the previous period county-level sugarcane deflated price; $P_{l,t-1}$ is the previous period county-level substitute prices vector;); $REVSC_{i,t-1}$ is the previous period county-level sugarcane gross revenue per hectare; $REV_{l,t-1}$ is the previous period county-level substitute uses gross revenue per hectare vector; $DMill_{i,t-1}$ is the previous period distance from county i to the nearest mill; $\nu_{i,j}$ is the county-level fixed effects; and μ_t is the year fixed effects. We assumed $\varepsilon_{i,j,t} \sim N(0, \sigma^2)$. By construction, the coefficients α and β are short-run elasticities, and long-run elasticities are given by $\beta_k/(1 - \alpha)$, where $k = 1, \ldots, 6$. These different specifications will improve the overall understanding of farmers' land allocation decisions.

To estimate the empirical models, a panel including 243 counties and 8 years (2006 to 2013), a strongly balanced panel, was used. The model was estimated using the command *xtreg* with fixed-effects and robust variance on software STATA IC 14.

2.4   Data

In the empirical land allocation model, sugarcane acreage is the dependent variable. The acreage comes from the LULC maps (section 2.2). Summary statistics for the data used to implement the empirical models including previous acreage, yield, price, and revenue are presented in Table 1. Crop yields and prices are from the IBGE (2016b). Cattle herd uses data from the IBGE (2016a) for the county-level herd; cattle price is from the CONAB at state-level for the period 2010 to 2013. The state price was interpolated to the county level using the same price gap recorded in the 2006 Agricultural Census (IBGE 2006).

Acreage data for Cerrado, Soybean double crop, Soybean single crop, pasture, and forest are from the LULC maps developed for this study. Among the land use considered, pasture is the one with the largest mean, and it is the only land use present in each county of the study area. Cerrado is the second largest land use, while Sugarcane is the smallest land use in the study area. Concerning the acreage data, it is important to say that a value of 0.000001 was added to classes with a value of zero to enable the use of log-transformation.

Given the assumption of farmers' profit maximization behavior, we include three sets of variables to capture the influence of gross revenue on the land allocation process. The Price dataset is from IBGE (2016b). Table 1 shows that sugarcane has the lowest price, almost one order of magnitude smaller than the price of corn

Table 1 Summary statistics for the dependent and explanatory variables

Variable (unit)	Mean	SD	Max	Min
Sugarcane (ha)	1772	4983	45,141	0
Cerrado (ha)	59,943	95,562	740,223	0
Soybean double crop (ha)	2478	12,080	220,516	0
Soybean single crop (ha)	11,026	26,749	354,159	0
Pasture (ha)	89,671	133,404	1,327,962	1866
Forest (ha)	7200	13,239	120,413	17
Yield sugarcane (tons/ha)	36.7	30.9	140.0	–
Yield corn (tons/ha)	4.2	1.8	12.0	–
Yield soy (tons/ha)	2	1.3	6.5	–
Cattle stockage (head/ha)	1.45	0.6	5.56	0.11
Price sugarcane (R$/ton)	42.7	38.1	430	–
Price corn (R$/ton)	302.4	99	701.2	–
Price soy (R$/ton)	393.3	295	1400	–
Price herd (R$/head)	1343	256	2445	0
Revenue sugarcane (R$/ha)	2060	1932	34,400	–
Revenue soybean (R$/ha)	1114	891	4490	–
Revenue corn (R$/ha)	1301	700	5077	–
Revenue cattle (R$/head)	1922	813	6616	–
Distance to mill (km)	72.4	46.2	241.0	1.38

and soybeans. Yield is also used to proxy gross revenue. Sugarcane presents an average yield that is one order of magnitude bigger than the yield of the other crops (Table 1). Revenue variables are calculated as the product of *Price x Yield*, given an approximation of the gross revenue per hectare; except for Cattle revenue which is calculated as *Price x Cattle Stockage*. This variable is a direct approximation to possible revenue.

Distance to mill is incorporated into the model to capture the influence of ethanol mills in the land allocation decision.

3   Results and Discussion

Given our interest on the effect of previous land use on farmers' acreage response, we start our analysis focusing on the acreage variables. First, we focus on sugarcane own response and distance to the mill. Second, we examine the relationship with natural vegetation. Third, we examine the relationship with other crops. Last, we focus on the relationship with pasture. The estimates for four models of acreage response using fixed-effects panel analysis are presented in Table 2. Before analyzing the results, it is important to note that Model 3 has fewer observations than the other models given the definition of revenue. This model represents the counties with presence of sugarcane while the other models are examining the whole study area. We will focus the discussion on the results of Models 3 and 4, while results of Models 1 and 2 are used as supporting arguments. We selected Model 4 because it represents the overall land allocation decision, while Model 3 is more representative of counties with sugarcane.

Sugarcane shows a positive relationship with its own lagged areas in all the specifications. The estimates in this model are short-run elasticities, thus sugarcane is elastic to its own areas. Areas with sugarcane in the previous year has a positive but inelastic response because all the coefficients are positive and smaller than 1. Also, the long-run elasticity is a function of the sugarcane acreage (lagged) (α), which results in a long-run elasticity (see section 2.3 for the definition of long-run elasticity). In the long-run sugarcane has an elastic response meaning that changes in the sugarcane area in the previous year prompts larger land allocation for sugarcane.

These results show that sugarcane has a spatial inertia in the short-run (Hausman 2012), however is elastic in the long-run. The spatial inertia can be understood given five factors of sugarcane production. First, sugarcane has a long-term production cycle, being economically viable for six crop-years (Goldemberg 2006). Second, sugarcane grows in a ratoon system which increases the cost to replace this crop before the end of the long-term production cycle (Nogueira et al. 2013). Third, the implementation of a sugarcane farm is a costly process encompassing slope correction, soil preparation, and plots systematization for mechanical harvest (Egeskog et al. 2016; Sparovek et al. 2007). Fourth, the machinery involved in the farming of this crop is specially designed for it which reduces farmers ability to produce other crops (Aguiar et al. 2011; Coelho et al. 2006; Van Den Wall Bake

Table 2 Acreage response of sugarcane in the states of Goiás and Mato Grosso do Sul

Variable	Model 1 (Eq 1) Short-run	Long-run	Model 2 (Eq 2) Short-run	Long-run	Model 3 (Eq 3) Short-run	Long-run	Model 4 (Eq 4) Short-run	Long-run
Sugarcane (ha)	0.579***	1.377	0.578***	1.372	0.532***	1.136	0.573***	1.342
Soybean single crop (ha)	−0.014	−0.034	−0.009	−0.020	−0.005	−0.011	−0.013	−0.030
Soybean double crop (ha)	0.023**	0.056	0.023**	0.056	0.008	0.018	0.029***	0.069
Pasture (ha)	0.672	1.599	0.570	1.351	−0.590	−1.261	0.576	1.348
Cerrado (ha)	0.151	0.360	0.328	0.777	0.065	0.140	0.377	0.883
Nearest mill (km)	−0.471***	−1.120	−0.464***	−1.100	−0.422***	−0.902	−0.446***	−1.045
Yield sugarcane (tons/ha)	0.011	0.026					0.116*	0.272
Yield corn (tons/ha)	0.029	0.069					−0.191	−0.448
Yield soybeans (tons/ha)	0.078	0.186					0.102	0.238
Cattle stockage (heads/ha)	0.086	0.205					0.058	0.136
Price sugarcane (R$/ton)			−0.009	−0.021			−0.106*	−0.249
Price soybeans (R$/ton)			0.012	0.028			0.039*	0.092
Price corn (R$/ton)			−0.005	−0.012			−0.027	−0.063
Price cattle herd (R$/head)			1.296**	3.075			0.973*	2.279
Revenue sugarcane (R$/ha)					0.027	0.057		
Revenue soybeans single (R$/ha)					0.055	0.117		
Revenue corn (R$/ha)					0.012	0.025		
Revenue cattle (R$/ha)					0.165	0.353		
Constant	−5.874		−15.726		7.303		−14.059	
N	1944		1936		986		1936	
Cluster	243		242		155		242	
R-squared (within)	0.5213		0.5218		0.4196		0.5218	

Note: *significant at 0.1, **significant at 0.05, ***significant at 0.01

et al. 2009). Fifth, the contracts between suppliers and ethanol mills are based on the production cycles instead of years (McKee et al. 2017; Sant'Anna et al. 2016a). All these factors combine to influence farmers to keep producing sugarcane until a drastic shock takes place. The shock can be positive, such as the increase in demand for ethanol in Brazil and new mills, or it can be negative, such as the decrease in the oil price and bankruptcy of ethanol mills.

One condition for this expansion is the presence of ethanol mills. Mills are the only destination of large-scale sugarcane production. The constraint on the spatial distribution of sugarcane is because of the degradation of the sugar content used to sugar and ethanol production. Once harvested, sugarcane needs to be processed in <72 h which results in an average radius of 45 km (Capaz et al. 2013; Neves et al. 1998). Therefore, the relation between mills and producing areas reinforces the spatial inertia of sugarcane. The estimate for the distance to the nearest mill shows that the acreage response is negative and inelastic, thus an increase in the distance results in a decrease in the acreage of sugarcane. This response is elastic in the long-run. For instance, when a new mill opens, it reduces the distance of surrounding farms thus stimulating the increase of sugarcane production. Moreover, mills do not start their operation crushing at full capacity. Mills usually need three to five harvest to optimize their operation and develop the ability to operate at full capacity thus increasing the demand for sugarcane. Therefore, the mill's learning curve contribute to explain the elasticity in the long-run.

The acreage model also examined the land dynamics of sugarcane and natural vegetation. This dynamic is explicitly modeled by incorporating the acreage of natural vegetation (Cerrado) as a competing use in the sugarcane allocation model. The results do not find Cerrado as a statistically significant factor in the study area. Notably, this finding corroborates the argument that sugarcane expansion is not being attracted by natural vegetation but rather to areas already converted to agricultural/pasture use (Adami et al. 2012; Aguiar et al. 2011; Ferreira Filho and Horridge 2014; Rudorff et al. 2010). This finding can be further understood when we consider the presence of the Sugarcane Agroecological Zoning which zoned out the conversion of natural vegetation to sugarcane expansion after 2008 (Manzatto et al. 2009).

A major concern related to biofuel production is the competition between the production of food-crop and fuel-crop. We investigate this competition by considering the two main row-crop productions as alternative uses in the farmers' decision processes. In the partial-adjustment models, land competition would appear as negative acreage response to acreage allocated to alternative uses indicating that an increase of food-crop would lead to a decrease of fuel-crop. However, the estimate for the Soybean double crop acreage is positive and significant (Models 1, 2, and 4) while for the Soybean single crop, it is negative but not significant. These results indicate that instead of competition (negative elasticity), intensification of row-crop production (Soybean double crop) is consistent with sugarcane expansion, thus these land uses presents signs of complementary (positive elasticity). This interpretation is supported by the cross-price response to soybeans, which is positive but inelastic indicating that an increase of 1% in the price of soybeans has positive change of 1% to the land allocated to sugarcane (significative in Model 4 and not significative in Model 2).

The conversion of pasture to sugarcane is another alternative of land use transition, such transition is very important in the Brazilian Cerrado given its large extend of degraded pastureland. This transition is the one favored by the Brazilian government because of its smaller impact on food production when compared to the conversion of cropland. Even though previous research has found that sugarcane expanded over pasture (Adami et al. 2012; Aguiar et al. 2011; Ferreira Filho and Horridge 2014; Rudorff et al. 2010), the estimate for Pasture acreage is not statistically significant (Table 2). Model 3 coefficient for pasture acreage is negative indicating the expected competition given the literature on sugarcane conversion of pastureland. The cross-price acreage response to Price of Cattle is significant and elastic, implying that an increase in the price of cattle positively impacts the allocation of land to sugarcane. The intensification of pasture and cattle stockage has been proposed as a source of 'new' agricultural land and an alternative to reduce deforestation in the Cerrado (Gil et al. 2015; Mann et al. 2014; Martha et al. 2012). The strong cross-price elasticity demonstrates the importance of cattle ranching on farmers' decisions. With the increase of cattle price, ranchers can obtain more revenue that in turn can be applied to the intensification of cattle ranching activity, thus freeing land for sugarcane. Therefore, this finding may support the claim of intensification of pasture as a source of land.

Considering sugarcane own-price elasticity, it is negative, ineslatic (close to zero) and statistically significant on Model 4. The negative elasticity is not expected indicating that a price reduction would lead to an area expansion. However, this can be seem as another confirmation of the spatial inertia. Given that farmers are already committed to sugarcane production, they can sustain their gross revenue by increasing area. Negative elasticity has also been found in other studies of sugarcane acreage response (Hausman 2012). When we consider the revenue (Model 3) the own-elasticity becomes positive but not significant. This change of sign may indicate that a different process is taking place among counties that are already growing sugarcane.

4 Conclusion

In conclusion, this research uses a time-series of land use maps, thus farmers' decisions over time can be modeled considering the relationship between sugarcane and the other main agricultural land use in the Cerrado portion of the states of Goiás and Mato Grosso do Sul. This research advances three points to better understand the land allocation to sugarcane. The first point focuses on the impact of the mills. Our results found evidence that distance to mills is negatively correlated with sugarcane acreage, thus mills play a role in the sugarcane expansion. The second point is concerned with the Cerrado vegetation. The results show no statistically significant correlation of Cerrado and sugarcane acreage. The third point looks at the land use competition. This research indicates the importance of cattle ranching and the intensification of grain production in understanding farmers' sugarcane acreage decisions in the states of Goiás and Mato Grosso do Sul. Furthermore, a new land use dynamic between sugarcane and grain production is found. The intensification from a single crop to a double crop is correlated with an increase of sugarcane acreage; thus, these land uses tend to be complementary. This finding lends support to the claim that intensification of grain production can release land for biofuel and/or sugar production. However, this result cannot support the claim that this land use change does not affect food production and price.

This study contributes to the development of policies aiming at improving the sustainability of sugarcane production in the Cerrado. For instance, by supporting the intensification of grain sector, the government is also supporting the production of sugarcane and reducing the competition between food and fuel production. On the other side, policies should continue to inhibit the conversion of natural vegetation to sugarcane as the Sugarcane Agroecological Zoning does. For the period considered, sugarcane acreage was not influenced by the amount of natural vegetation in the specfic country, a result that must continue in the future to improve sugarcane's sustainability. Additionally, policies can focus on the location of new ethanol mills. By directing the sitting of mills to areas with the smallest competition with other land uses such as pasture, the government can influence the sugarcane expansion to more desired paths, such as the conversion of pasture to sugarcane.

Acknowledgments The National Science Foundation supported this work through the grant [NSF BCS 1227451 – "Collaborative Research: Land Change in the Cerrado: Ethanol and Sugar Cane Expansion at the Farm and Industry Scale"]. The authors thank the collaboration of Kansas Applied Remote Sensing Program, especially Jude Kastens, Christopher Bishop and J. Christopher Brown. The authors also thank the comments received at the American Association of Geographers and at the Conference of Latin Americanist Geographers meetings. We thank the anonymous reviewers for their helpful comments.

References

Adami M, Rudorff BFT, Freitas RM, Aguiar DA, Sugawara LM, Mello MP (2012) Remote sensing time series to evaluate direct land use change of recent expanded sugarcane crop in Brazil. Sustainability 4(12):574–585. https://doi.org/10.3390/su4040574

Aguiar DA, Rudorff BFT, Silva WF, Adami M, Mello MP (2011) Remote sensing images in support of environmental protocol: monitoring the sugarcane harvest in São Paulo state, Brazil. Remote Sens 3(12):2682–2703. https://doi.org/10.3390/rs3122682

Alkimim A, Sparovek G, Clarke KC (2015) Converting Brazil's pastures to cropland: an alternative way to meet sugarcane demand and to spare forestlands. Appl Geogr 62:75–84. https://doi.org/10.1016/j.apgeog.2015.04.008

Arvor D, Meirelles M, Dubreuil V, Bégué A, Shimabukuro YE (2012) Analyzing the agricultural transition in Mato Grosso, Brazil, using satellite-derived indices. Appl Geogr 32(2):702–713. https://doi.org/10.1016/j.apgeog.2011.08.007

Barretto AGOP, Berndes G, Sparovek G, Wirsenius S (2013) Agricultural intensification in Brazil and its effects on land-use patterns: an analysis of the 1975-2006 period. Glob Chang Biol 19(6):1804–1815. https://doi.org/10.1111/gcb.12174

Bergtold JS, Caldas MM, Sant'anna AC, Granco G, Rickenbrode V (2017) Indirect land use change from ethanol production: the case of sugarcane expansion at the farm level on the Brazilian Cerrado. J Land Use Sci:1–15. https://doi.org/10.1080/1747423X.2017.1354937

Bergtold JS, Fewell J, Williams J (2014) Farmers' willingness to produce alternative cellulosic biofuel Feedstocks under contract in Kansas using stated choice experiments. Bioenergy Res 7(3):876–884. https://doi.org/10.1007/s12155-014-9425-9

Brannstrom C, Filippi AM (2008) Remote classification of *Cerrado* (savanna) and agricultural land covers in northeastern Brazil. Geocarto Int 23(2):109–134. https://doi.org/10.1080/10106040701596767

Brannstrom C, Jepson W, Filippi AM, Redo D, Xu Z, Ganesh S (2008) Land change in the Brazilian savanna (Cerrado), 1986–2002: comparative analysis and implications for land-use policy. Land Use Policy 25(4):579–595. https://doi.org/10.1016/j.landusepol.2007.11.008

Brazil M (2015) Mapeamento do Uso e Cobertura do Cerrado: Projeto Terra Class Cerrado 2013. Ministério do Meio Ambiente (MMA), Brasília, p 67

Breiman L (2001) Random forests. Mach Learn 45(1):5–32. https://doi.org/10.1023/A:1010933404324

Brown JC, Kastens JH, Coutinho AC, Victoria D d C, Bishop CR (2013) Classifying multiyear agricultural land use data from Mato Grosso using time-series MODIS vegetation index data. Remote Sens Environ 130:39–50. https://doi.org/10.1016/j.rse.2012.11.009

Brown J, Jepson W, Kastens J, Wardlow B, Lomas J, Price K (2007) Multitemporal, moderate-spatial-resolution remote sensing of modern agricultural production and land modification in the Brazilian Amazon. GI Sci Remote Sensing 44(2):117–148. https://doi.org/10.2747/1548-1603.44.2.117

Bustamante MC, Nobre C, Smeraldi R, Aguiar AD, Barioni L, Ferreira L et al (2012) Estimating greenhouse gas emissions from cattle raising in Brazil. Clim Chang 115(3–4):559–577. https://doi.org/10.1007/s10584-012-0443-3

Caldas MM, Bergtold JS, Peterson JM, Graves RW, Earnhart D, Gong S et al (2014) Factors affecting farmers' willingness to grow alternative biofuel feedstocks across Kansas. Biomass Bioenergy 66(0):223–231. https://doi.org/10.1016/j.biombioe.2014.04.009

Caldas MM, Granco G, Bishop C et al (2017) Effects of sugarcane ethanol expansion in the Brazilian cerrado. In: Qin Z, Mishra U, Hastings A (eds) Bioenergy and land use change. John Wiley & Sons, Inc., pp 39–51

Caldas MM, Sanderson MR, Mather M, Daniels MD, Bergtold JS, Aistrup J et al (2015) Opinion: Endogenizing culture in sustainability science research and policy. Proc Natl Acad Sci 112(27):8157–8159. https://doi.org/10.1073/pnas.1510010112

Capaz RS, Carvalho VSB, Nogueira LAH (2013) Impact of mechanization and previous burning reduction on GHG emissions of sugarcane harvesting operations in Brazil. Special Issue on Advances in sustainable biofuel production and use - XIX International Symposium on Alcohol Fuels - ISAF 102(0):220–228. https://doi.org/10.1016/j.apenergy.2012.09.049

Carpentier A, Letort E (2014) Multicrop production models with multinomial Logit acreage shares. Environ Resour Econ 59(4):537–559. https://doi.org/10.1007/s10640-013-9748-6

Carvalho FMV, De Marco Júnior P, Ferreira LG (2009) The Cerrado into-pieces: habitat fragmentation as a function of landscape use in the savannas of Central Brazil. Biol Conserv 142(7):1392–1403. https://doi.org/10.1016/j.biocon.2009.01.031

Clark ML, Aide TM, Grau HR, Riner G (2010) A scalable approach to mapping annual land cover at 250 m using MODIS time series data: a case study in the dry Chaco ecoregion of South America. Remote Sens Environ 114(11):2816–2832. https://doi.org/10.1016/j.rse.2010.07.001

Coelho ST, Goldemberg J, Lucon O, Guardabassi P (2006) Brazilian sugarcane ethanol: lessons learned. Energy Sustain Dev 10(2):26–39. https://doi.org/10.1016/S0973-0826(08)60529-3

Cohn AS, Mosnier A, Havlík P, Valin H, Herrero M, Schmid E et al (2014) Cattle ranching intensification in Brazil can reduce global greenhouse gas emissions by sparing land from deforestation. Proc Natl Acad Sci 111(20):7236–7241. https://doi.org/10.1073/pnas.1307163111

de CONAB CNA (2016) Safras: Séries históricas - Cana-de-Açúcar. www.conab.gov.br/OlalaCMS/uploads/arquivos/16_04_18_14_58_56_canaseriehist.xls

Egeskog A, Barretto A, Berndes G, Freitas F, Holmén M, Sparovek G, Torén J (2016) Actions and opinions of Brazilian farmers who shift to sugarcane—an interview-based assessment with discussion of implications for land-use change. Land Use Policy 57:594–604. https://doi.org/10.1016/j.landusepol.2016.06.022

Ferreira ME, Ferreira LG, Latrubesse EM, Miziara F (2016) Considerations about the land use and conversion trends in the savanna environments of Central Brazil under a geomorphological perspective. J Land Use Sci 11(1):33–47. https://doi.org/10.1080/1747423X.2013.845613

Ferreira ME, Ferreira LG, Miziara F, Soares-Filho BS (2013) Modeling landscape dynamics in the central Brazilian savanna biome: future scenarios and perspectives for conservation. J Land Use Sci 8(4):403–421. https://doi.org/10.1080/1747423X.2012.675363

Ferreira NC, Ferreira LG, Huete AR, Ferreira ME (2007) An operational deforestation mapping system using MODIS data and spatial context analysis. Int J Remote Sens 28(1):47–62. https://doi.org/10.1080/01431160600835861

Ferreira Filho JB d S, Horridge M (2014) Ethanol expansion and indirect land use change in Brazil. Land Use Policy 36:595–604. https://doi.org/10.1016/j.landusepol.2013.10.015

Gil J, Siebold M, Berger T (2015) Adoption and development of integrated crop–livestock–forestry systems in Mato Grosso, Brazil. Agric Ecosyst Environ 199:394–406. https://doi.org/10.1016/j.agee.2014.10.008

Goldemberg J (2006) The ethanol program in Brazil. Environ Res Lett 1(1):14008. https://doi.org/10.1088/1748-9326/1/1/014008

Goldemberg J, Mello FFC, Cerri CEP, Davies CA, Cerri CC (2014) Meeting the global demand for biofuels in 2021 through sustainable land use change policy. Energy Policy 69(0):14–18. https://doi.org/10.1016/j.enpol.2014.02.008

Granco G (2017) Land change dynamics in the Brazilian Cerrado the interaction of biofuels, markets, and biodiversity. Kansas State University, Manhattan, Kan

Granco G, Caldas MM, Bergtold JS, Sant'Anna AC (2017) Exploring the policy and social factors fueling the expansion and shift of sugarcane production in the Brazilian Cerrado. Geo J 82(1):63–80. https://doi.org/10.1007/s10708-015-9666-y

Granco G, Sant'Anna AC, Bergtold JS, Caldas MM (2018) Factors influencing ethanol mill location in a new sugarcane producing region in Brazil. Biomass and Bioenergy 111:125–133. https://doi.org/10.1016/j.biombioe.2018.02.001

Haile MG, Kalkuhl M, von Braun J (2016) Worldwide acreage and yield response to international price change and volatility: a dynamic panel data analysis for wheat, Rice, corn, and soybeans. Am J Agric Econ 98(1):172–190. https://doi.org/10.1093/ajae/aav013

Hansen MC, Potapov PV, Moore R, Hancher M, Turubanova SA, Tyukavina A et al (2013) High-resolution global maps of 21st-century Forest cover change. Science 342(6160):850–853. https://doi.org/10.1126/science.1244693

Hausman C (2012) Biofuels and land use change: sugarcane and soybean acreage response in Brazil. Environ Resour Econ 51(2):163–187. https://doi.org/10.1007/s10640-011-9493-7

Hendricks NP, Smith A, Sumner DA (2014) Crop supply dynamics and the illusion of partial adjustment. Am J Agric Econ 96(5):1469–1491. https://doi.org/10.1093/ajae/aau024

Hennessy DA (2006) On monoculture and the structure of crop rotations. Am J Agric Econ 88(4):900–914. https://doi.org/10.1111/j.1467-8276.2006.00905.x

Horta Nogueira LA, Moreira JR, Schuchardt U, Goldemberg J (2013) The rationality of biofuels. Energy Policy 61(0):595–598. https://doi.org/10.1016/j.enpol.2013.05.112

IEA - Instituto de Economia Agrícola (2014) Protocolo Agroambiental do Setor Sucroenérgetico Paulista : Dados consolidados das safras 2007/2008 a 2013/2014. http://www.iea.sp.gov.br/out/LerTexto.php?codTexto=13453

Instituto Brasileiro de Geografia e Estatística (2006) Censo agropecuário 2006: resultados preliminares. IBGE

Instituto Brasileiro de Geografia e Estatística, I (2016a). Sistema IBGE de Recuperação Automática - SIDRA: Produção Agrícola Municipal. http://www.sidra.ibge.gov.br/bda/acervo/acervo9.asp?e=c&p=PA&z=t&o=11

Instituto Brasileiro de Geografia e Estatística, I (2016b). Sistema IBGE de Recuperação Automática - SIDRA: Pesquisa Pecuaria Municipal. http://www.sidra.ibge.gov.br/bda/acervo/acervo9.asp?e=c&p=PP&z=t&o=24

Irwin EG, Geoghegan J (2001) Theory, data, methods: developing spatially explicit economic models of land use change. Agric Ecosyst Environ 85(1–3):7–23. https://doi.org/10.1016/S0167-8809(01)00200-6

Jepson W (2006) Private agricultural colonization on a Brazilian frontier, 1970–1980. J Hist Geogr 32(4):839–863. https://doi.org/10.1016/j.jhg.2004.12.019

Jepson W, Brannstrom C, Filippi A (2010) Access regimes and regional land change in the Brazilian Cerrado, 1972–2002. Ann Assoc Am Geogr 100(1):87–111. https://doi.org/10.1080/00045600903378960

Kastens JH, Brown JC, Coutinho AC, Bishop CR, Esquerdo JCDM (2017) Soy moratorium impacts on soybean and deforestation dynamics in Mato Grosso, Brazil. PLoS One 12(4):e0176168. https://doi.org/10.1371/journal.pone.0176168

Klink CA, Machado RB (2005) Conservation of the Brazilian Cerrado. Conserv Biol 19(3):707–713. https://doi.org/10.1111/j.1523-1739.2005.00702.x

Leal MRLV, Horta Nogueira LA, Cortez LAB (2013) Land demand for ethanol production. Special Issue on Advances in sustainable biofuel production and use - XIX International Symposium on Alcohol Fuels - ISAF 102(0):266–271. https://doi.org/10.1016/j.apenergy.2012.09.037

Mann ML, Kaufmann RK, Bauer DM, Gopal S, Nomack M, Womack JY et al (2014) Pasture conversion and competitive cattle rents in the Amazon. Ecol Econ 97(0):182–190. https://doi.org/10.1016/j.ecolecon.2013.11.014

Manzatto CV, Assad ED, Bacca JFM, Zaroni MJ, Pereira SEM (2009) Zoneamento agroecológico da cana-de-açúcar. Documentos, Rio de Janeiro

Martha GB, Alves E, Contini E (2012) Land-saving approaches and beef production growth in Brazil. Agric Syst 110(0):173–177. https://doi.org/10.1016/j.agsy.2012.03.001

McKee B, Sant'Anna AC, Bergtold JS et al (2017) TRUST ME! Examining the contractual relationships between sugarcane producers and mills in the Cerrado. Rev Extensão e Estud Rurais 6:98. doi:https://doi.org/10.18540/rever62201798-117

Ministério do Meio Ambiente (2008) Projeto de Conservação e Utilização Sustentável da Diversidade Biológica Brasileira - Probio I. http://mapas.mma.gov.br/mapas/aplic/probio/datadownload.htm

Ministério do Meio Ambiente (2014) PPCerrado – Plano de Ação para prevenção e controle do desmatamento e das queimadas no Cerrado:2ª fase (2014-2015). MMA, Brasilia. http://www.mma.gov.br/images/arquivos/florestas/controle_e_prevencao/PPCerrado/PPCerrado_2fase.pdf. Accessed 19 September 2016

Müller H, Rufin P, Griffiths P, Barros Siqueira AJ, Hostert P (2015) Mining dense Landsat time series for separating cropland and pasture in a heterogeneous Brazilian savanna landscape. Remote Sens Environ 156:490–499. https://doi.org/10.1016/j.rse.2014.10.014

Myers N, Mittermeier RA, Mittermeier CG, da Fonseca GAB, Kent J (2000) Biodiversity hotspots for conservation priorities. Nature 403(6772):853–858. https://doi.org/10.1038/35002501

Neves MF, Waack RS, Marino MK (1998) Sistema agroindustrial da cana-de-açúcar: caracterização das transações entre empresas de insumos, produtores de cana e usinas. In Congresso Da Sociedade Brasileira De Economia E Sociologia Rural–Sober (Vol. 36)

Newberry D (2014) Why are there cattle in the conservation area? Social barriers to biofuel governance in Brazil. Geoforum 54(0):306–314. https://doi.org/10.1016/j.geoforum.2013.08.011

Peterson JM, Caldas MM, Bergtold JS, Sturm BS, Graves RW, Earnhart D et al (2014) Economic linkages to changing landscapes. Environ Manag 53(1):55–66. https://doi.org/10.1007/s00267-013-0116-7

Rada N (2013) Assessing Brazil's Cerrado agricultural miracle. Food Policy 38:146–155. https://doi.org/10.1016/j.foodpol.2012.11.002

Rodrigues DMT, Miziara F (2008) Expansão da fronteira agrícola: a intensificação da pecuária bovina no Estado de Goiás. Pesquisa Agropecuária Tropical (Agricultural Research in the Tropics) 38(1):14–20. www.revistas.ufg.br/pat/article/view/3613

Rudorff BFT, de Aguiar DA, da Silva WF, Sugawara LM, Adami M, Moreira MA (2010) Studies on the rapid expansion of sugarcane for ethanol production in São Paulo state (Brazil) using Landsat data. Remote Sens 2(4):1057–1076. https://doi.org/10.3390/rs2041057

Saddiq M, Fayaz M, Hussain Z, Shahab M, Ullah I (2014) Acreage response of sugarcane to price and non price factors in Khyber Pakhtunkhwa. Inter J Food Agr Econ 2(3):121–128. http://www.foodandagriculturejournal.com/vol2.no3.pp121.pdf

Sano EE, Rosa R, Brito JLS, Ferreira LG (2008) Mapeamento semidetalhado do uso da terra do Bioma Cerrado. Pesq Agrop Brasileira 43(1):153–156. https://doi.org/10.1590/S0100-204X2008000100020

Sano EE, Rosa R, Brito JLS, Ferreira LG (2010) Land cover mapping of the tropical savanna region in Brazil. Environ Monit Assess 166(1–4):113–124. https://doi.org/10.1007/s10661-009-0988-4

Sant'Anna AC, Granco G, Bergtold J, Caldas MM, Xia T, Masi P et al (2016a) Os desafios da expansão da cana-de-açúcar: a percepção de produtores e arrendatários de terras em Goiás e Mato Grosso do Sul. In: dos Santos GR (ed) Quarenta anos de etanol em larga escala no Brasil: desafios, crises e perspectivas. IPEA, Brasilia, p 315. http://www.ipea.gov.br/portal/images/stories/PDFs/livros/livros/160315_livro_quarenta_anos_etanol.pdf

Sant'Anna AC, Shanoyan A, Bergtold JS, Caldas MM, Granco G (2016b) Ethanol and sugarcane expansion in Brazil: what is fueling the ethanol industry? Inter Food Agr Manag Rev 19(4):163–182. https://doi.org/10.22434/IFAMR2015.0195

Secchi S, Kurkalova L, Gassman PW, Hart C (2011) Land use change in a biofuels hotspot: the case of Iowa, USA. Biomass Bioenergy 35(6):2391–2400. https://doi.org/10.1016/j.biombioe.2010.08.047

Shikida PFA (2013) Expansão canavieira no Centro-Oeste: limites e potencialidades. Revista de Política Agrícola 22(2):122–137. https://seer.sede.embrapa.br/index.php/RPA/article/view/312. Accessed 1 June 2017

Silva AA, Miziara F (2011) Avanço do setor sucroalcooleiro e expansão da fronteira agrícola em Goiás. Pesquisa Agropecuária Tropical (Agricultural Research in the Tropics) 41(3). https://doi.org/10.5216/pat. v41i3. 11054

Sparovek G, Berndes G, Egeskog A, Mazzaro de Freitas FL, Gustafsson S, Hansson J (2007) Sugarcane ethanol production in Brazil: an expansion model sensitive to socioeconomic and environmental concerns. Biofuels Bioproducts & Biorefining-Biofpr 1(4):270–282. https://doi.org/10.1002/bbb.31

Strassburg BBN, Brooks T, Feltran-Barbieri R, Iribarrem A, Crouzeilles R, Loyola R et al (2017) Moment of truth for the Cerrado hotspot. Nature Ecology Evolut 1(4):99. https://doi.org/10.1038/s41559-017-0099

Strassburg BBN, Latawiec AE, Barioni LG, Nobre CA, da Silva VP, Valentim JF et al (2014) When enough should be enough: improving the use of current agricultural lands could meet production demands and spare natural habitats in Brazil. Glob Environ Chang 28(0):84–97. https://doi.org/10.1016/j.gloenvcha.2014.06.001

Van den Wall Bake JD, Junginger M, Faaij A, Poot T, Walter A (2009) Explaining the experience curve: cost reductions of Brazilian ethanol from sugarcane. Biomass Bioenergy 33(4):644–658. https://doi.org/10.1016/j.biombioe.2008.10.006

Walker R, Browder J, Arima E, Simmons C, Pereira R, Caldas M et al (2009) Ranching and the new global range: Amazônia in the 21st century. Geoforum 40(5):732–745. https://doi.org/10.1016/j.geoforum.2008.10.009

Walter A, Galdos MV, Scarpare FV, Leal MRLV, Seabra JEA, da Cunha MP et al (2014) Brazilian sugarcane ethanol: developments so far and challenges for the future. Wiley Interdisciplinary Rev: Energy and Environ 3(1):70–92. https://doi.org/10.1002/wene.87

Wardlow BD, Egbert SL (2008) Large-area crop mapping using time-series MODIS 250 m NDVI data: an assessment for the U.S. central Great Plains. Remote Sens Environ 112(3):1096–1116. https://doi.org/10.1016/j.rse.2007.07.019

Wardlow BD, Kastens JH, Egbert SL (2006) Using USDA crop progress data for the evaluation of Greenup onset date calculated from MODIS 250-meter data. Photogramm Eng Remote Sens 72(11):1225–1234

Wardlow B, Egbert S, Kastens J (2007) Analysis of time-series MODIS 250 m vegetation index data for crop classification in the U.S. central Great Plains. Remote Sens Environ 108(3):290–310. https://doi.org/10.1016/j.rse.2006.11.021

Wu J, Brorsen BW (1995) The impact of government programs and land characteristics on cropping patterns. Canadian J Agr Econ/Revue canadienne d'agroeconomie 43(1):87–104. https://doi.org/10.1111/j.1744-7976.1995.tb00109.x

Growing Switchgrass in the Corn Belt: Barriers and Drivers from an Iowa Survey

Sarah Varble and Silvia Secchi

Abstract While agriculture has dramatically increased the production of crops for energy generation, there has been limited growing of dedicated perennial crops for liquid fuel or electricity production. Adoption of dedicated perennials can be the first step in the transformation from unsustainable, energy intensive agricultural production to a system that is resilient to climate change, environmentally sustainable and financially stable for farmers. We focus on the perennial switchgrass because it is a native species and there is evidence of its favorable agronomic and environmental profile under a wide range of growing conditions. However, since switchgrass is a novel, perennial crop, there are challenges for farmers who want to grow it. This paper examines which factors are significant predictors in the interest of farmers in producing switchgrass through the analysis of the results of a survey completed by farmers in the Clear Creek watershed in rural Iowa. Knowledge of switchgrass and production on highly erodible land are both highly correlated with interest in growing switchgrass. In addition, long-term contracts with bio-refineries would help farmers decrease adoption risk. Our results can help establish policies that could influence farmers to shift production from annual crops to perennial native biomass energy crops. Switchgrass production has the potential to move agriculture from a contributor to climate change into a sector that mitigates climate change via reduction in energy-intensive inputs, such as fertilizers, production of renewable fuels, and sequestration of carbon in the soils.

Keywords Switchgrass adoption · Sustainability · Mitigation · Adaptation · Risk aversion · Innovation · Corn Belt

S. Varble
Southern Illinois University-Carbondale, Carbondale, IL, USA
e-mail: svarble@siu.edu

S. Secchi (✉)
University of Iowa, Iowa, IA, USA
e-mail: silvia-secchi@uiowa.edu

© Springer International Publishing AG, part of Springer Nature 2018
R. Li, A. Monti (eds.), *Land Allocation for Biomass Crops*,
https://doi.org/10.1007/978-3-319-74536-7_7

125

1 Introduction

As climatic variations increase with rising carbon dioxide (IPCC 2012), agro-ecosystems will become more strained due to increasing threats from droughts, floods, and extreme temperatures, and this will affect the agricultural-based communities that sector supports (IPCC 2014). The United States (US) is not immune from the impacts of extreme climate events. The 2012 drought is just one example of how vulnerable agro-ecosystems are to climate change events. It caused over $14.2 billion in crop loss (Shields 2013). The US Corn Belt is a region crucial to the production of food, fuel and fiber for both the US and global markets. However, in the region, many regulating ecosystem services such as water filtration, carbon sequestration, biodiversity, soil stabilization and wildlife habitat have been lost due to production methods which can cause soil erosion, water quality and quantity degradation, and loss of diversity (Tilman et al. 2002). These environmental impacts have led to a decrease in the sustainability of the agro-ecosystem in social, economic, and environmental terms, which are the three measurements of sustainability as defined by the Triple Bottom Line (Elkington 1998). Additionally, climate change, because of its impacts on extreme events, could reduce the socio-ecological resilience of the Corn Belt agro-ecosystem, that is, the level of disturbance that the system can withstand while retaining its function and identity (Carpenter et al. 2001). Switchgrass could provide a viable adaptation to climate change (Smith and Olesen 2010)

Adaptations, or changes made in response to actual or predicted climatic events to reduce their impacts (IPCC 2013) take many forms; from small, low risk changes that require fewer inputs (such as capital and infrastructure) to large changes that necessitate a paradigm shift and significant capital outlay (Park et al. 2012). The adoption of adaptation activities can reduce some risks faced by farmers and agro-ecosystems. One such activity, the addition of switchgrass to a production regime, can be used as a transformative adaptation to assist farmers as they face increasing climate change threats. Switchgrass is a native to North American, and it can help increase both sustainability and resilience of the agro-ecosystem because it requires fewer fertilizer and pesticide applications, reduces fertilizer runoff and soil erosion, increases carbon sequestration, and is drought and flood tolerant due to its deep root system (Cope et al. 2011; Groom et al. 2008; McLaughlin and Walsh 1998). Switchgrass is considered a practical alternative to annual row crops particularly for less productive and highly erodible lands (Johnson et al. 2010).

From an economic perspective, adding switchgrass to a farmer's production portfolio has the potential to decrease their reliance on one or two annual row crops such as corn and soybeans. Agricultural activities are increasingly being utilized for climate change mitigation via carbon sequestration, both above and below ground, and through the production of crops grown for biofuels, which displace fossil fuel. Research has demonstrated that switchgrass is capable of both carbon sequestration, through its intensive biomass both above and underground, and the displacement of fossil fuels (Groom et al. 2008; Schmer et al. 2008; Ma et al. 2000). Thus, since switchgrass can be combusted and used for heat and electricity or made into ethanol, it can provide local job opportunities and climate change mitigation (Smith and Olesen 2010). This would offer both economic and social benefits to rural areas.

Switchgrass has the potential to produce in excess of 500% more energy in ethanol than what is necessary to grow and harvest it (Schmer et al. 2008). These types of mitigation practices are particularly important because they are relatively low cost compared to other non-agricultural mitigation options (Strengers et al. 2008), and could have substantial co-benefits (McKinley et al. 2011). The adoption of agricultural mitigation practices is not without challenges, however, since carbon sequestration in agriculture is dependent on continued management, and can shift carbon-emitting activities elsewhere (Smith et al. 2007).

The challenges in using switchgrass as an adaptation and mitigation tool are related to its novelty, the fact that it requires a long term commitment for farmers (stands can last over 15 years) (Wen et al. 2009), and the risk averse and profit maximization inclinations exhibited by farmers (Just 1975). Because switchgrass is a novel crop, there are many unknown variables in its adoption, such as equipment and labor needs, profitability, and opportunity costs. Adopting switchgrass would increase initial variability in income and profit, both of which are avoided by farmers (Babcock and Shogren 1995). Additionally, the recent record high corn prices have motivated farmers to put more land in corn production (Secchi et al. 2008). These two forces merge to stall switchgrass adoption. Additional barriers to adoption include the absence of markets, land tenure, and time conflicts. Previous work on farmers' willingness to adopt bioenergy crops suggested a variety of barriers to adoption, including socioeconomic characteristics (Skevas et al. 2014; Jensen et al. 2007; Gedikoglu 2012), land suitability (Skevas et al. 2014), contractual arrangements (Bergtold et al. 2014), and information (Villamil et al. 2012). We include all these issues in our analysis.

The Diffusion of Innovation Theory, which is used to explain the adoption process of new innovations, can also be used to explain some barriers to adoption of new technologies and products, and posits that access to information and knowledge of the innovation is the initial step in the adoption process (Rogers 2003). The source of the information is crucial to adoption (Nicholas and Durham 2012), and studies have found that if the information is disseminated by a close friend, or someone who is similar in nature to the potential adopter, the likelihood and speed of adoption increases (Wejnert 2002). This study will examine whether information sources play a role in the dissemination of knowledge about switchgrass and how that affects interest in switchgrass adoption.

Our study took place in watershed in Iowa, a state where corn and soybean production is intense and has resulted in negative environmental externalities such as nitrogen loading in ground and surface water, channelization of waterways, and decreased biodiversity. The choice of a watershed as the study unit was driven by the importance of water quality as a key sustainability indicator. The watershed is the appropriate spatial unit to confront sustainability challenges linked to water

(Druschke 2013). To examine farmer and landowner interest in growing switchgrass, as well as barriers and drivers of adoption, we conducted a survey of farmers and landowners in a small watershed in Iowa.

2 Materials and Methods

2.1 Study Area

The survey was conducted in the Clear Creek watershed in eastern Iowa. Clear Creek is 25 miles in length and is a tributary to the Iowa River, which feeds into the Mississippi River. The watershed includes 65,000 acres across two counties: Iowa and Johnson. Cities within the watershed include Iowa City, Coralville, and North Liberty. However, the watershed is primarily agricultural, with approximately 65% of the land cover in corn, soybean, or pasture/alfalfa cultivation (Druschke and Secchi 2014). This watershed is unique and ideal for switchgrass production because nearly 57% of the cropland in the Clear Creek watershed is highly erodible, or land that has the capacity to erode at excessive rates (U.S. Department of Agriculture 2012). Forty percent of the cropland is both highly erodible and highly suitable for corn production, which differs from other areas in state, where only 6% of cropland is highly erodible (see Fig. 1 and Table 1). This is unusual for Iowa, but not for the Corn Belt as a whole (Wilhelm et al. 2004).

In 2004, the Environmental Protection Agency listed Clear Creek on its 303(d) Impaired Waters list (Druschke and Secchi 2014), and since then, much work has been done to improve the water quality of Clear Creek. Though Clear Creek itself is no longer on EPA's impaired waters 303(d) list, the watershed is part of the Iowa river basin, which suffers from severe water quality issues – American Rivers chose it as one of the country's most endangered rivers in 2007. The mandate for the implementation of a Total Maximum Daily Load (TMDL), which is a regulatory tool used under the Clean Water Act in the US to address water pollution (Kerr 2014), the consequent availability of Federal funding, and local concerns over water quality resulted in the creation of a watershed management group to educate residents and farmers living within the watershed.

2.2 Description of Survey and Sample

A mail survey was distributed in April 2010 to 998 rural landowners and farmers in the Clear Creek watershed asking about agricultural and conservation practices relevant to the Corn Belt region in general, adoption of best management practices and adaptation. Before the survey was mailed to the respondents, it was pretested with a group of experts to increase its reliability and validity. To obtain the names and addresses of the potential respondents, extensive on the ground preparatory

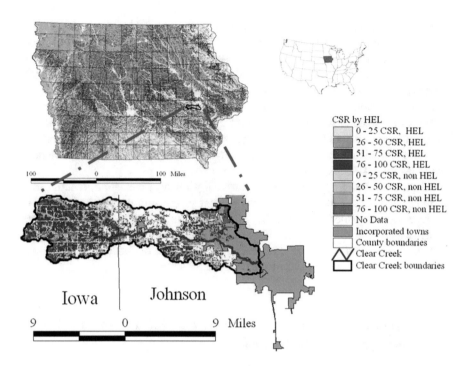

Fig. 1 Corn Suitability Rating (CSR) for Highly Erodible Land (HEL) and Non-HEL in the Clear Creek Watershed. (Sources: Iowa Cooperative Soil Survey 2003, USDA NASS Research and Development Division 2017)

Table 1 Crop land based on Corn Suitability Rating (CSR) and highly erodible land (HEL) designation in Clear Creek Watershed, Iowa, and for the entire state of Iowa

	Clear creek		Iowa	
	Acres	**Percentage of cropland**	**Acres**	**Percentage of cropland**
CSR 76–100, non HEL	10,452	28.66%	13,829,125	58.45%
CSR 51–75, non HEL	5122	14.04%	8,346,070	35.28%
CSR 26–50, non HEL	151	0.41%	123,742	0.52%
CSR 0–25, non HEL	54	0.15%	3787	0.02%
CSR 76–100, HEL	1054	2.89%	1644	0.01%
CSR 51–75, HEL	13,806	37.85%	117,192	0.50%
CSR 26–50, HEL	4563	12.51%	348,664	1.47%
CSR 0–25, HEL	1273	3.49%	889,127	3.76%
	36,474	**Area cropped in 2012**	23,659,351	**Area cropped in 2012**
	66,134	**Overall watershed area**	35,993,064	**Overall Iowa area**
	55.15%	**Percentage cropland**	65.73%	**Percentage cropland**

Sources: Iowa Cooperative Soil Survey 2003, USDA NASS Research and Development Division 2017.

work was conducted - geographic information system (GIS) county parcel data was used to identify rural parcels and exclude urban areas. The list of land parcels was cross-referenced against a list of names and addresses provided by the Farm Service Agency office for each county. The survey, which was sent to all 998 landowners and active farmers in the watershed that we could identify, was 16 page long and comprised of 62 questions. The survey included six sections: personal information, watershed conservation, information sources, farm characteristics, on farm conservation practices, and growing crops for biofuels. Of the surveys mailed, 31 were undeliverable, and 397 were returned completed, after two reminder mailings, for a 41% response rate.

The survey sample had very good variability for age, education, income from farming, and acres farmed. Fifty-eight percent of the sample was male, and the average age of respondents was 62, which follows the trend of aging farmers. Ages ranged from 25 to 97 (see Table 2 for a summary of all of the descriptive statistics). Our data was consistent with the US Census of Agriculture for farmer demographics. The data was analyzed using SPSS software.

2.3 Interest in Planting Switchgrass: Theoretical Foundations and Hypothesis Construction

We examined the interest in adopting switchgrass as a crop in relation to structural socio-economic characteristics of farmers, their knowledge of the crop (and where it came from), and risk-related considerations. As noted in the introduction, these are all variables that have been found to be critical in the adoption of innovation and adaptation in the agricultural sector.

Socio-demographic and "descriptive" variables have shown to be predictive in both the adoption of innovations overall, and specifically with the adoption of switchgrass. Gedikoglu (2012) found that younger respondents and respondents with more education were more willing to grow crops specifically for use in energy production (Gedikoglu 2012). Jensen et al. (2007) found similar results. In addition, Jensen et al. found that farm profitability per acre was a significant predictor of interest in growing switchgrass, however the percent of off-farm income was not found to be a significant predictor. Age, education, and disposable income are all variables that have shown to be significant in the adoption of other innovations as well, including technological and medical innovations. Younger, more educated adopters with higher incomes are more apt to take risks and adopt innovations more rapidly than older, less educated people with lower incomes (Rogers 2003; Wejnert 2002). Additionally, if a producer has more off-farm income, there is a greater likelihood that they are less reliant on on-farm income to remain financially stable and are more flexible in their farm decisions.

Knowledge of the innovation is the first phase in the Rogers' Theory of Innovation decision period. Knowledge can reduce uncertainty by providing information about the innovation and thus reduce risk (Rogers 2003). Therefore, it can be

Table 2 Descriptive statistics of variables

	N	Minimum	Maximum	Mean	Std. Deviation
Demographic info					
Age	306	25	97	63	14
Gender	313	0	1	0	0
Education	310	1	6	3	2
Gross income from agriculture	263	1	6	2	1
% Household income from farming	270	0%	76–100%	26–50%	1
Information sources					
NRCS	313	0	1	0	0
FSA	313	0	1	0	0
COEXT	313	0	1	0	1
ISU	313	0	1	0	0
AGROCHEM DEAL	313	0	1	0	0
NEIGHFRIENDS	313	0	1	0	0
SCDC	313	0	1	0	0
AG INSTRUCT	313	0	1	0	0
MACH DEAL	313	0	1	0	0
Land ownership and production					
Total acres owned	274	0	3000	201	341
Tillable acres owned	256	0	2900	170	321
Acres in CRP	230	0	446	17	50
Total acres farmed	230	0	3040	192	421
HEL acres farmed	211	0	2500	113	311
Switchgrass variables					
SGINTEREST	222	0	1	0	0
KNOWLEDGE	227	1	10	3	2
Minimum net profit/acre required	88	50	1000	204	144
Acres converted to SG with minimum net profit	136	0	1000	48	104
Acres converted to SG with 50% higher net profit	130	0	1000	65	155

Note: Totals are different because some questions were not answered. *NRCS* Natural Resources Conservation Service, *FSA* Farm Service Agency, *COEXT* County Extension, *ISU* Iowa State University, *SCDC* Soil Conservation District Commissioners

inferred that farmers and land owners will be more interested in growing switchgrass if they are aware of switchgrass characteristics, the environmental and financial benefits that it can provide, and the resilience it can add to agro-ecosystems. Based on the Diffusion of Innovation Theory that knowledge of switchgrass is a significant predictor of adoption, the source of information that a farmer uses to obtain conservation information will also be important (Nicholas and Durham 2012; Baumgart-Getz et al. 2012). Farmers and land owners must rely on their trusted sources to obtain credible information and advice about land management (Sagor and Becker 2014; Varble et al. 2015). Trust and legitimacy are important aspects among stakeholders receiving information from experts (Sabatier et al. 2005; Ostrom 1994; Ostrom et al. 1999), thus, the use of trusted experts as knowledge sources for switchgrass will not only decrease uncertainty, but also increase willingness to adopt new innovations. Tucker and Napier (2002) found that the leading sources of information about soil and water conservation were the Farm Service Agency and the Natural Resource Conservation Service (government agricultural agencies) and agro-chemical dealers (Tucker and Napier 2002). However, because the purpose of their job is to sell seeds, chemicals, and other agricultural products to enhance corn and soybean production, it is unlikely that agro-chemical dealers will be distributing information about the benefits of switchgrass. It is more likely that County Extension agents and university specialists (in Iowa, employed by Iowa State University) who are focused on sustainable agriculture practices will provide information about switchgrass production to farmers and landowners.

Risk considerations are important for farmers when deciding whether to adopt new practices or technologies. As mentioned, farmers tend to be risk averse (Just 1975), therefore, there must be some type of motivation in place to adopt innovations such as switchgrass, such as financial motivation. It is improbable that farmers will grow switchgrass on prime farmland that produces high corn and soybean yields. In 2007, Babcock et al. found switchgrass would need to be able to bring in more than $437 per acre, for a profit of $250/acre, in order to compete with corn and soybeans (Babcock et al. 2007). The price farmers would receive for switchgrass would likely not be able to compete with the price for corn and soybeans. In addition, the lack of an established market for switchgrass is also seen as unfavorable in terms of risk reduction.

Switchgrass' suitability for use on highly erodible lands and land suitable for enrollment in the Conservation Reserve Program (CRP) makes it possible that farmers and landowners will see it as an alternative to enrolling land in the Conservation Reserve Program. The CRP was established in 1985 by the US Department of Agriculture as a way to set aside highly erodible and less productive, fragile land in order to decrease erosion, and it targeted highly erodible land (HEL). Since 1985, the CRP program has broadened its goals to a wider Environmental Benefit Index, which includes wildlife habitat and water pollution, so not all CRP land is HEL.

HEL is an ideal location for switchgrass production because it generally has lower production yields of corn and soybeans, and planting annual crops on HEL can have higher environmental impacts. However, in the Clear Creek Watershed, there is a large quantity of land that is categorized as highly erodible that also has a high corn suitability rating (CSR) (see Fig. 1), which means that while it is sloped, it is still very fertile (Iowa Cooperative Soil Survey 2003; USDA NASS Research and Development Division 2017). Land owners who enroll in CRP are paid a competitive market rental rate to maintain a long-term native perennial vegetative cover (Cowan et al. 2009). If farmers believe that switchgrass is an alternative to CRP, the profit per acre required would be comparable to the CRP rental rate. In Iowa, the average CRP rental rate in Johnson and Iowa counties is $130.50/acre. Farmers who receive direct payments through the USDA must also perform more conservation practices when farming HEL, which makes this type of land more difficult to farm. Additionally, Cope et al. (2011) found that survey respondents in a different Corn Belt area were most interested in adopting switchgrass on marginal land. The lack of established markets for switchgrass increases the risk for farmers and acts as a deterrent to adoption. For a farmer to adopt switchgrass, other financial incentives (besides profit per acre) would be required to lower the risk associated with growing and harvesting the switchgrass. There are several methods through which could be used to achieve this outcome, including contract type and length, subsidies, and insurance.

Using this literature as our theoretical basis, we hypothesize:

H1a: Younger, more educated farmers/land owners will show more interest in growing switchgrass.

H1b: Percent of off-farm income will be a significant predictor in interest in growing switchgrass.

H2: Farmers and land owners who are knowledgeable about switchgrass will be more interested in growing it.

H3: Farmers and landowners who use government agricultural agencies, County Extension agents and university specialists as information sources will be more knowledgeable about switchgrass.

H4: The farmers and land owners who are interested in growing switchgrass will grow it on land such as HEL or CRP land.

2.4   Variable Construction

The dependent variable, interest in growing switchgrass, was constructed from two of the survey questions: "Would you be interested in growing switchgrass on land you own and farm?", and "Would you be interested in engaging in long term rental contracts to rent some of your land to somebody who wants to grow switchgrass or to rent land for switchgrass production yourself?". The combination of these two questions addressed the fact that many farmers both own and rent land, and many land owners only lease their land to be farmed and do not farm it themselves. Thus, all options would be covered. They were given the answer choices "yes" coded "1", "no" coded "0", and "I do not own any land", coded "0". The answers to both of these questions were combined into a single variable (SGINTEREST). If they answered yes to either question, it was coded "1".

The level of switchgrass knowledge was measured using multiple questions. The first, "How knowledgeable are you about the harvesting and marketing of switchgrass?" asked respondents to state their knowledge level on a Likert-type scale ranging from 1 (not at all knowledgeable) to 5 (very knowledgeable). Seventy-one percent of respondents rated themselves as "not at all knowledgeable", and only 2% said they were "very knowledgeable". The second set of questions tested the actual knowledge of respondents on their environmental awareness of switchgrass (Fig. 2).

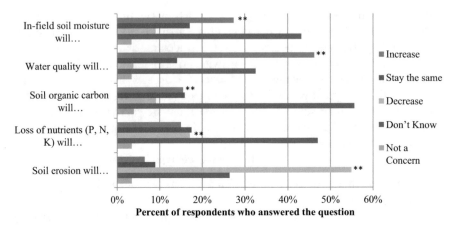

Fig. 2 Responses to "If grain crops are replaced with switchgrass for biofuel..." **correct answer

For the logistic regression analysis, if the respondent answered an environmental awareness statement correctly, it was coded as "1", otherwise "0". All of the coded statements were added together and combined with the switchgrass production and marketing statement to create an overall switchgrass knowledge measure (KNOWLEDGE).

To obtain sources of information, respondents were asked "Who is/are your main source(s) of information on conservation issues (please check all that apply)?" The sources listed in the survey instrument included: Natural Resource Conservation Service; Farm Service Agency (combined into "federal government agricultural agency); County Extension Service; university specialists; local agribusiness dealers; neighbors and friends; Soil Conservation District Commissioners; and vocational agriculture instructors. If the respondent marked that they used a specific information source, it was coded "1", and otherwise it was "0".

To determine whether the respondents perceive switchgrass as a secondary crop to produce on less productive or fragile land (CRP or HEL), the respondents were asked a number of questions about land ownership and farming, including the number of acres enrolled in CRP, tillable acres owned, acres farmed, and acres of HEL farmed. They were also asked about profit: "What is the minimum net profit per acre you would need to get in order to consider growing switchgrass?" Additionally, respondents were asked the number of acres they would plant in switchgrass if they could get that net profit, and the number of acres they would plant if they could get a net profit of 50% more.

The preferred method of risk reduction in terms of financing and contracts was determined through the following statement: "To sell switchgrass to bio-refineries, I would require.... a long term contract (3–5 years) with the bio-refinery; a short term contract (1–2 years) with the bio-refinery; government subsidies; special insurance to lower risk; bank financing; and using a co-op to handle delivery arrangements." Respondents were given the following answer choices: strongly disagree, disagree, neutral, agree, strongly agree or unsure (Table 3).

Table 3 Responses to "To sell switchgrass to bio-refineries I would require . . . "

	Interest in Switch-grass?	Unsure	Strongly Disagree	Disagree	Neutral	Agree	Strongly Agree	Total
Long term contract with the bio-refinery	No	70	2	3	22	16	8	121
	Yes	29	0	0	6	30	21	86
Short-term contract with the bio-refinery	No	72	4	9	19	14	1	119
	Yes	32	10	9	17	11	5	84
Government subsidies	No	65	2	0	27	20	5	119
	Yes	30	4	6	13	22	9	84
Special insurance to lower risk	No	70	1	4	23	16	4	118
	Yes	31	3	6	18	23	2	83
Bank financing	No	70	3	8	24	13	1	119
	Yes	29	6	10	25	11	3	84
Using a co-op to handle delivery arrangements	No	67	3	3	27	20	0	120
	Yes	33	5	1	11	30	5	85

Standard statistical texts available in SPSS were implemented to assess the performance of the model. To determine goodness of fit, since there is no definitive test used for logistic regressions, we used a variety of tests. The Hosmer and Lemeshow test measures the fit between the model's expected results and the actual data, giving a measure of whether the model is a good predictor. The Cox and Snell and Nagelkerke are pseudo R Squares. They compare the log likelihood of the model with that of the null model with only the constant. The Nagelkerke test is an adjusted Cox and Snell - it allows for the pseudo R Square to get to 1.

3   Results

Hypotheses 1a, 1b and 2 were tested using a logistic regression with SGINTEREST as the dependent variable. Forty-two percent of respondents (n = 93) who answered the question stated that they would be interested in growing switchgrass. The model omnibus fit test was significant (χ^2 (7) = 32.107, p = 0.000) indicating that the model used has explanatory power. The Hosmer and Lemeshow test was insignificant (χ^2 (9) = 5.152, p = 0.741), signifying that logistic regression was a good fit for the data. The model explained between 21.6% (Cox and Snell R Square) and 29% (Nagelkerke R Square) of the variance. Based on the results from this test, unlike previous studies, we did not find age, education, or percent income from farming to be significant predictors (Jensen et al. 2007; Gedikoglu

Table 4 Responses to "Harvesting switchgrass from fields will..."

	Strongly Disagree	Disagree	Neutral	Agree	Strongly Agree	Unsure	Total
Increase equipment needs	2 (0.8%)	3 (1.3%)	24 (10%)	88 (36.7%)	38 (15.8%)	85 (35.4%)	240
Increase the need for custom baling	2 (0.8%)	6 (2.5%)	30 (12.4%)	83 (34.4%)	31 (12.9%)	89 (36.9%)	241
Increase the need for joint ownership of balers	4 (1.7%)	8 (3.4%)	52 (21.9%)	58 (24.5%)	10 (4.2%)	105 (44.3%)	237
Increase the need for specialized management to ensure the quality of switchgrass	2 (0.8%)	9 (3.8%)	40 (16.7%)	66 (27.6%)	23 (9.6%)	99 (41.4%)	239
Increase the need for clean storage areas	1 (0.4%)	0 (0%)	33 (13.8%)	84 (35.1%)	23 (9.6%)	98 (41%)	239

2012). KNOWLEDGE was the only significant variable ($F = 11.05$, $p = 0.001$) for predicting interest in switchgrass production in the logistic regression, supporting hypothesis 2.

After performing a correlation between all of the independent and dependent variables, we found SGINTEREST to be strongly positively correlated with education and negatively correlated with age and percent income from farming, which lends some support to H1a. There was a strong positive correlation between KNOWLEDGE and SGINTEREST, confirming the results of the logistic regression (Tables 3 and 4).

Most respondents (63.6%) used a federal government agricultural agency as a source of information, followed by the County Extension Service (48.7%), friends and neighbors (45.5%). Our results were similar to those found by Tucker and Napier (2002) with government agricultural agencies being popular sources (Tucker and Napier 2002). Seventy-six percent of respondents used 3 sources or less for information about conservation (see Fig. 3). To test H3, a correlation was run between sources of information and KNOWLEDGE, and the results showed that KNOWLEDGE was negatively correlated (at the 0.05 level) with the county extension service and neighbors and friends. A t-test confirmed these results, which means that the respondents who use these sources for conservation information know less about switchgrass than other respondents. In addition, the respondents who used the federal government agriculture agencies and Iowa State University were more knowledgeable about switchgrass. This shows partial support for H3.

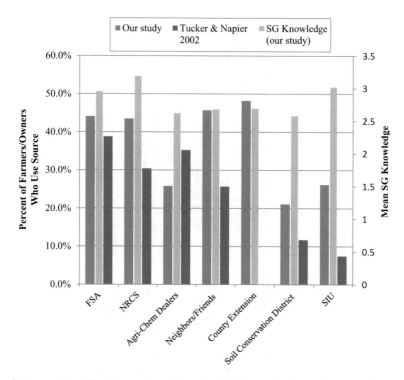

Fig. 3 Sources Used for Conservation Information and Corresponding Switchgrass Knowledge, ranging from a 1 (low knowledge) to 5 (very knowledgeable), obtaining combining all Likert scale questions on switchgrass knowledge

The required stated profit per acre ranged from $50/acre to $1000/acre, and the average profit required was $203/acre, which falls between both the CRP rental rates ($130.50/acre) and the $250 required return over cost calculated by Babcock et al. (2007).

There was a strong significant relationship between the number of HEL acres farmed and the number of acres respondents would be willing to plant if they received their base profit ($r = 0.503$, $p = 0.000$). In addition, our results showed that the farmers who farm HEL in the Clear Creek watershed are price responsive, as they would be willing to plant more acres if they received more money per acre. CRP owners were not as willing to plant switchgrass if they received the minimum amount they required, however, they were also price responsive, and would plant more acres in switchgrass if they received more money. These results show strong support for H4 for highly erodible land and partial support for CRP land (see Table 5).

Nearly 60% of all respondents who were not interested in growing switchgrass answered "unsure" when asked whether they would require a contract, subsidy, insurance, or financing to sell switchgrass, which reiterates the importance of knowledge in the adoption of new technologies. Among the respondents who were

Table 5 Correlations for Hypothesis 4

	HEL acres	CRP acres	Tillable acres owned	Total Acres Farmed	Required SG Net Profit	# of SG Acres with Net Profit	# of SG Acres with 50% higher net profit
HEL acres farmed	1	0.080	0.234**	0.880**	0.018	0.503**	0.396**
Enrolled CRP acres	0.080	1	0.341**	0.029	−0.017	0.179	0.187*
Tillable acres owned	0.234**	0.341**	1	0.249**	0.106	0.412**	0.691**
Total acres farmed	0.880**	0.029	0.249**	1	0.026	0.413**	0.324**
Required net profit	0.018	−0.017	0.106	0.026	1	0.020	0.063
Number of switchgrass acres with net profit	0.503**	0.179	0.412**	0.413**	0.020	1	0.890**
Number of switchgrass acres with 50% more than net profit	0.396**	0.187*	0.691**	0.324**	0.063	0.890**	1

**Correlation is significant at the 0.01 level (2-tailed); *Correlation is significant at the 0.05 level (2-tailed)

interested in growing switchgrass, long-term contracts were the preferred method of risk reduction, as 59% of respondents agreed or strongly agreed with this choice. Using a co-op to handle delivery arrangements was also popular with respondents interested in switchgrass production (41% agreed with this statement).

4 Discussion and Conclusion

Based on the results of our study and previous literature, we know that farmer characteristics, such as age and education, do play a role in farmer and land owner interest in switchgrass production; however the most significant factor in switchgrass adoption is knowledge, which confirms the Diffusion of Innovation Theory. We found that most farmers and landowners in the study relied on the county extension office, neighbors and friends, and government agencies for information about conservation, yet, the respondents who used the agribusiness dealers and family and friends as sources of information knew the least about switchgrass. Since we know which sources farmers and land owners trust for information, these should be the prime outlets for farmer and land owner education about switchgrass. In addition, since most of the respondents were not aware of harvest or financial requirements, education would help with marketing switchgrass to farmers and land owners (Table 4).

The high correlation between the number of highly erodible acres farmed and the number of acres a respondent is willing to put into switchgrass production likely tells the story of switchgrass adoption in the US. As we already stated, it is unlikely that farmers in the US will replace corn with switchgrass on highly productive land, but switchgrass is a more likely crop for HEL as it has deep root systems, requires minimal tillage, and minimal inputs. However, until markets are created for switchgrass and subsidies are provided, it is highly unlikely that any type of mainstream adoption will occur.

The fact that switchgrass is a perennial and has an extensive root system makes it ideal for carbon sequestration and nutrient runoff absorption (Lee and Mickelson 1998; Liebig et al. 2008). In the US, mitigation markets and markets for ecosystem services have been slow to launch, but if an ecosystem services market were created to reduce the hypoxia in the Gulf of Mexico allowing for the participation of non-point sources in the Corn Belt, adoption of switchgrass could play a key role. However, until that occurs, or more biorefineries are opened, the adoption of switchgrass, even in highly erodible lands or land suited for the Conservation Reserve Program, is unlikely.

Table 6 Summary of hypotheses with results

	Hypothesis	Results
H1a	Younger, more educated farmers/landowners will show more interest in growing switchgrass	Supported in correlation, not supported in regression
H1b	Percent of off-farm income will be a significant predictor in interest in growing switchgrass	Supported in correlation, not supported in regression
H2	Farmers and land owners who are knowledgeable about switchgrass will be more interested in growing it	Supported
H3	Farmers and landowners who use government agricultural agencies (NRCS and FSA), county extension, and university specialists for information sources will be more knowledgeable about switchgrass	Supported for NRCS, FSA, university specialists
H4	The farmers and land owners who are interested in growing switchgrass will grow it on land such as HEL or CRP land	Supported for HEL, partial support for CRP

Our results Table 6 show that once bio-refineries are in place, the key factor to switchgrass adoption is a long-term contract. In addition, we also found that farmers are price responsive. Respondents stated that they would be willing to plant more acres in switchgrass if they received higher profits. In context with the previous result, this suggests that farmers would consider a new technology such as switchgrass, which offers environmental benefits but also production risks, for more fragile portions of their farms, if adequately compensated. This is important for policymakers because they can provide more financial incentives and subsidies in order to reduce the risk of switchgrass adoption, and farmers are more likely to adopt switchgrass where it would be more environmentally beneficial. However, if corn prices continue to increase, farmers will require elevated subsidies as an incentive to continue switchgrass production instead of replacing it with corn on HEL or marginal land.

Switchgrass is unlikely to be adopted on a broad scale in the Corn Belt, however, it can be very valuable environmentally, socially, and economically if planted on fragile lands such as HEL. If adopted in these areas, switchgrass can work as both a climate change adaptation and mitigation technique for farmers by providing soil and water conservation, replacing fossil fuels, and sequestering carbon. Its worth goes beyond the economic value it can provide agricultural producers.

While our results are geographically limited to a small watershed, they are indicative of farmers' attitudes in the US Corn Belt, which is a particularly important corn production region – the US is responsible for over a third of global corn production (USDA Office of the Chief Economist 2017), and the Corn Belt accounts for 90% of it (Schnitkey 2013). This region is also critically important in causing – and thus potentially reducing – the hypoxic zone in the Gulf of Mexico (Scavia et al. 2003; Turner and Rabalais 2003; Rabotyagov et al. 2014), which is the second largest in the world (Scavia et al. 2017). The Clear Creek watershed contains a large percentage of highly erodible land. As mentioned, this is not the norm in Iowa, but, in the Corn Belt as a whole, farming on highly erodible land is a problem, thus our results have a potentially much larger significance. However, to expand the results and gain a better perspective of the opinions of farmers and landowners, it would be beneficial to replicate the survey across multiple watersheds and states.

Acknowledgements This work was funded in part by the National Science Foundation Cyber-enabled Discovery and Innovation Award 0835607 and Coupled Natural Human Systems Award 1114978. The views expressed are solely the authors' own. We thank the editors and two reviewers for their constructive comments and suggestions.

References

Babcock BA, Gassman PW, Jha M, Kling CL (2007) Adoption subsidies and environmental impacts of alternative energy crops. Iowa State University, Center for Agricultural and Rural Development

Babcock BA, Shogren JF (1995) The cost of agricultural production risk. Agric Econ 12(2):141–150

Baumgart-Getz A, Prokopy LS, Floress K (2012) Why farmers adopt best management practice in the United States: a meta-analysis of the adoption literature. J Environ Manag 96(1):17–25. https://doi.org/10.1016/j.jenvman.2011.10.006

Bergtold JS, Fewell J, Williams J (2014) Farmers' willingness to produce alternative cellulosic biofuel feedstocks under contract in Kansas using stated choice experiments (journal article). Bioenergy Res 7(3):876–884. https://doi.org/10.1007/s12155-014-9425-9

Carpenter S, Walker B, Anderies JM, Abel N (2001) From metaphor to measurement: resilience of what to what? (journal article). Ecosystems 4(8):765–781. https://doi.org/10.1007/s10021-001-0045-9

Cope MA, McLafferty S, Rhoads BL (2011) Farmer attitudes toward production of perennial energy grasses in east Central Illinois: implications for community-based decision making. Ann Assoc Am Geogr 101(4):852–862. https://doi.org/10.1080/00045608.2011.575320

Cowan T, Johnson R, Stubbs M (2009) Conservation Provisions of the 2008 Farm Bill (Updated January 13, 2009). Congressional Research Service Report for Congress, Order Code RL34557

Druschke CG (2013) Watershed as common-place: communicating for conservation at the watershed scale. Environmental Communication 7(1):80–96. https://doi.org/10.1080/17524032.2012.749295

Druschke CG, Secchi S (2014) The impact of gender on agricultural conservation knowledge and attitudes in an Iowa watershed. J Soil Water Conserv 69(2):95–106. https://doi.org/10.2489/jswc.69.2.95

Elkington J (1998) Cannibals with forks: the triple bottom line of 21st century business. Capstone Publishing Limited, Oxford

Gedikoglu H (2012) 'Impact of off-farm employment on farmers' willingness to grow switchgrass and Miscanthus' *2012 Annual Meeting, February 4–7, 2012*. Southern Agricultural Economics Association, Birmingham

Groom MJ, Gray EM, Townsend PA (2008) Biofuels and biodiversity: principles for creating better policies for biofuel production. Conserv Biol 22(3):602–609. https://doi.org/10.1111/j.1523-1739.2007.00879.x

Iowa Cooperative Soil Survey (2003) Iowa Soil Properties and Interpretation Database

IPCC (2012) Summary for policymakers. In: Field CB, Barros V, Stocker TF, Qin D, Dokken DJ, Ebi KL (eds) Managing the risks of extreme events and disasters to advance climate change adaptation. A Special Report of Working Groups I and II of the Intergovernmental Panel on Climate Change, Cambridge/New York, pp 3–21

IPCC (2013) In: Stocker TF, Qin D, Plattner G-K, Tignor M, Allen SK, Boschung J (eds) Climate change 2013: the physical science basis. Contribution of working group I to the fifth assessment report of the intergovernmental panel on climate change. Cambridge, United Kingdom, New York, p 1535

IPCC (2014) In: Field CB, Barros VR, Dokken DJ, Mach KJ, Mastrandrea MD, Bilir TE, Chatterjee M, Ebi KL, Estrada YO, Genova RC, Girma B, Kissel ES, Levy AN, MacCracken S, Mastrandrea PR, White LL (eds) Climate change 2014: impacts, adaptation, and vulnerability. Part a: global and sectoral aspects. Contribution of working group II to the fifth assessment report of the intergovernmental panel on climate change. Cambridge University Press Cambridge, New York

Jensen K, Clark CD, Ellis P, English B, Menard J, Walsh M et al (2007) Farmer willingness to grow switchgrass for energy production. Biomass Bioenergy 31(11):773–781

Johnson JMF, Karlen DL, Andrews SS (2010) Conservation considerations for sustainable bioenergy feedstock production: if, what, where, and how much? J Soil Water Conserv 65(4):88A–91A. https://doi.org/10.2489/jswc.65.4.88A

Just RE (1975) Risk aversion under profit maximization. Am J Agric Econ 57(2):347–352. http://www.jstor.org/stable/1238513

Kerr L (2014) Compelling a nutrient pollution solution: how nutrient pollution litigation is redefining cooperative federalism under the clean water act. Environmental Law Lewis & Clark Northwestern School of Law 44(4):1219–1255

Lee, Isenhart, T., Schultz, R., & Mickelson, S. (1998). Nutrient and sediment removal by switchgrass and cool-season grass filter strips in Central Iowa, USA. Agrofor Syst, 44(2–3), 121–132

Liebig MA, Schmer MR, Vogel KP, Mitchell RB (2008) Soil carbon storage by switchgrass grown for bioenergy. Bioenergy Res 1(3–4):215–222

Ma Z, Wood CW, Bransby DI (2000) Soil management impacts on soil carbon sequestration by switchgrass. Biomass Bioenergy 18(6):469–477. https://doi.org/10.1016/S0961-9534(00)00013-1

McKinley DC, Ryan MG, Birdsey RA, Giardina CP, Harmon ME, Heath LS et al (2011) A synthesis of current knowledge on forests and carbon storage in the United States. Ecol Appl 21(6):1902–1924

McLaughlin SB, Walsh ME (1998) Evaluating environmental consequences of producing herbaceous crops for bioenergy. Biomass Bioenergy 14(4):317–324. https://doi.org/10.1016/S0961-9534(97)10066-6

Nicholas KA, Durham WH (2012) Farm-scale adaptation and vulnerability to environmental stresses: insights from winegrowing in northern California. Glob Environ Chang 22(2):483–494. https://doi.org/10.1016/j.gloenvcha.2012.01.001

Ostrom E (1994) Constituting social capital and collective action. Journal of Theoretical Politics 6(4):527–562. https://doi.org/10.1177/0951692894006004006

Ostrom E, Burger J, Field CB, Norgaard RB, Policansky D (1999) Revisiting the commons: local lessons, global challenges. Science 284(5412):278–282. https://doi.org/10.1126/science.284.5412.278

Park SE, Marshall NA, Jakku E, Dowd AM, Howden SM, Mendham E et al (2012) Informing adaptation responses to climate change through theories of transformation. Glob Environ Chang 22(1):115–126. https://doi.org/10.1016/j.gloenvcha.2011.10.003

Rabotyagov SS, Campbell TD, White M, Arnold JG, Atwood J, Norfleet ML et al (2014) Cost-effective targeting of conservation investments to reduce the northern Gulf of Mexico hypoxic zone. Proc Natl Acad Sci 111(52):18530–18535. https://doi.org/10.1073/pnas.1405837111

Rogers EM (2003) Diffusion of innovations. Free Press, New York

Sabatier F, Lubell M, Trachtenberg Z, Vedlitz A, Matlock M (2005) Swimming upstream: collaborative approaches to watershed management. The MIT Press, Cambridge, MA

Sagor ES, Becker DR (2014) Personal networks and private forestry in Minnesota. J Environ Manag 132(0):145–154. https://doi.org/10.1016/j.jenvman.2013.11.001

Scavia D, Bertani I, Obenour DR, Turner RE, Forrest DR, Katin A (2017) Ensemble modeling informs hypoxia management in the northern Gulf of Mexico. Proc Natl Acad Sci 114(33):8823–8828. https://doi.org/10.1073/pnas.1705293114

Scavia D, Rabalais NN, Turner RE, Justic D, Wiseman WJ (2003) Predicting the response of Gulf of Mexico hypoxia to variations in Mississippi River nitrogen load. Limnol Oceanogr 48(3):951–956. <Go to ISI>://000182982300001

Schmer MR, Vogel KP, Mitchell RB, Perrin RK (2008) Net energy of cellulosic ethanol from switchgrass (article). Proc Natl Acad Sci U S A 105(2):464–469. <Go to ISI>://000252551100015

Schnitkey G (2013) Concentration of Corn and Soybean Production in the U.S. Urbana-Champaign: Department of Agricultural and Consumer Economics, University of Illinois at Urbana-Champaign. Available at: http://farmdocdaily.illinois.edu/2013/07/concentration-corn-soybean-production.html

Secchi S, Tyndall J, Schulte LA, Asbjornsen H (2008) High crop prices and conservation. J Soil Water Conserv 63(3):68A–73A. https://doi.org/10.2489/jswc.63.3.68A

Shields D (2013) 'Federal Crop Insurance: Background' C. R Service December 12:2013

Skevas T, Swinton SM, Hayden NJ (2014) What type of landowner would supply marginal land for energy crops? *Biomass and Bioenergy* 67(supplement C):252–259. https://doi.org/10.1016/j.biombioe.2014.05.011

Smith P, Martino D, Cai Z, Gwary D, Janzen H, Kumar P et al (2007) Policy and technological constraints to implementation of greenhouse gas mitigation options in agriculture. Agric Ecosyst Environ 118(1–4):6–28. https://doi.org/10.1016/j.agee.2006.06.006

Smith P, Olesen JE (2010) Synergies between the mitigation of, and adaptation to, climate change in agriculture. J Agric Sci 148(05):543–552. https://doi.org/10.1017/S0021859610000341

Strengers BJ, Van Minnen JG, Eickhout B (2008) The role of carbon plantations in mitigating climate change: potentials and costs. Clim Chang 88(3–4):343–366

Tilman C,K, Matson P, Naylor R, Polasky S (2002) Agricultural sustainability and intensive production practices. Nature 418:671–677

Tucker M, Napier TL (2002) Preferred sources and channels of soil and water conservation information among farmers in three midwestern US watersheds. Agric Ecosyst Environ 92(2–3):297–313. https://doi.org/10.1016/S0167-8809(01)00293-6

Turner RE, Rabalais NN (2003) Linking landscape and water quality in the Mississippi River basin for 200 years. Bioscience 53(6):563–572. https://doi.org/10.1641/0006-3568(2003)053[0563:llawqi]2.0.co;2

U.S. Department of Agriculture (2012) Highly erodible land conservation and wetland conservation compliance fact sheet. N. a. FSA, Washington, D.C.

USDA NASS Research and Development Division (2017) Cropland Data Layer. Available at: http://www.nass.usda.gov/research/Cropland/SARS1a.htm

USDA Office of the Chief Economist, W. A. O. B (2017) Prepared by the interagency agricultural projections committee, 'USDA agricultural projections to 2026. Long-term projections report OCE-2017-1', Washington D.C. Available at: http://www.usda.gov/oce/commodity/projections/USDAAgriculturalProjections2022.pdf

Varble S, Secchi S, Druschke CG (2015) An examination of growing trends in land tenure and conservation practice adoption: results from a farmer survey in Iowa. Environ Manag:1–13. https://doi.org/10.1007/s00267-015-0619-5

Villamil MB, Alexander M, Silvis AH, Gray ME (2012) Producer perceptions and information needs regarding their adoption of bioenergy crops. Renew Sust Energ Rev 16(6):3604–3612. https://doi.org/10.1016/j.rser.2012.03.033

Wejnert B (2002) Integrating models of diffusion of innovations: a conceptual framework. Sociology 28(1):297

Wen Z, Ignosh J, Parrish D, Stowe J, Jones B (2009) Identifying farmers' interest in growing switchgrass for bioenergy in southern Virginia. J Ext 47(5)

Wilhelm WW, Johnson JMF, Hatfield JL, Voorhees WB, Linden DR (2004) Crop and soil productivity response to corn residue removal. Agron J 96(1):1–17. https://doi.org/10.2134/agronj2004.1000

Impact of Stover Collection on Iowa Land Use

Lyubov A. Kurkalova and Dat Q. Tran

Abstract The study evaluates land use impacts of corn stover markets for the state of Iowa. To tie land use decisions to their economic basis, we use an economic model to simulate profit-maximizing choices of crop-tillage rotations and stover collection, and evaluate the impacts of the stover collection restrictions imposed on the land of lower productivity, as defined by the land with Corn Suitability Rating below 80. We find that stover collection is likely to lead to substantial shifts in rotations favoring continuous corn at stover prices above \$50/ton. This crop rotation shift is accompanied by the changes in tillage rotations favoring both continuous conventional tillage and, to a lesser extent, continuous conservation tillage. The crop-rotation impacts of stover markets differ substantially between the restricted and unrestricted stover markets. This finding illustrates the importance of differentiating among the cropland of alternative soil quality when assessing the impacts of corn stover markets.

Silvia Secchi's assistance with data is gratefully acknowledged. The authors acknowledge partial support from U.S. Department of Agriculture (USDA) National Institute of Food and Agriculture, Agriculture and Food Research Initiative award No. 2016-67024-24755, and from the National Science Foundation (NSF) Bioenergy Center for Research Excellence in Science and Technology, award No. 1242152. The views expressed by the authors do not represent the official policy of USDA or NSF.

L. A. Kurkalova (✉)
Departments of Economics and Energy and Environmental Systems, North Carolina A&T State University, Greensboro, NC, USA
e-mail: lakurkal@ncat.edu

D. Q. Tran
Department of Agricultural Economics and Agribusiness, University of Arkansas, Fayetteville, AR, USA

1   Introduction

Advancements in cellulosic ethanol production technologies are expected to lead to the establishment of viable markets for corn residues (stover), which are comprised of corn stalks, cobs, and leaves left in the field after grain harvest. Recent research agrees that a large, viable market for stover is likely to significantly alter the profitability of corn relative to other traditional row crops and cropping patterns (Sarica and Tyner 2013; Dodder et al. 2015; USDOE 2016). However, large-scale analyses that commonly use relatively low-spatial-resolution models provide only limited insights on regional impacts. As soil and climatic conditions, cropping patterns, and farming practices differ across the U.S., the impacts of stover markets differ substantially between the states (Egbendewe-Mondzozo et al. 2011; Archer and Johnson 2012; Sesmero and Gramig 2013; Chen and Li 2016). Present study contributes to the literature on regional assessments of stover markets by evaluating the potential land use impacts for the state of Iowa.

Being a major U.S. corn producer, Iowa has been a subject of the economics of corn stover research (Kurkalova et al. 2010; Tyndall et al. 2011; Elobeid et al. 2013; Archer et al. 2014). Building on the models of Kurkalova et al. (2010) and Elobeid et al. (2013), we extend previous work by evaluating the impact of soil preservation restrictions on stover collection using a more realistic economic model of land use.

The land use model is extended in two important directions. First, we consider interactions between crop rotations and tillage. The crop rotation aspect is important because predominantly large fraction of Iowa cropland is in corn-soybean (CS) rotation (corn being yearly alternated with soybeans), with the rest of the land almost exclusively in continuous corn (CC) (corn planted every year) (Stern et al. 2008; Secchi et al. 2011b; Plourde et al. 2013). However, most previous economic analyses either focused on continuous corn (Archer et al. 2014), or have ignored crop rotations (USDOE 2016). Kurkalova et al. (2010) and Elobeid et al. (2013) model crop rotations, but under an assumption that tillage systems do not differ within any given crop rotation. Here we allow for more realistic choices, where tillage systems could alternate within crop rotations. Recent research suggests that such alternation is a wide-spread practice in Iowa (Kurkalova and Tran 2017).

Secondly, we relax the restriction that has been commonly imposed in previous analyses of Iowa crop production on the highly erodible land (HEL).[1] Kurkalova et al. (2010) and Secchi et al. (2011b) assumed that HEL is only farmed using no-till (NT), which is the tillage system that disturbs soils the least. That restriction is commonly imposed because the HEL designation requires farmers to implement conservation plans to remain eligible for payments from Federal agricultural programs (Claassen et al. 2014). However, monitoring of conservation compliance is

---

[1] USDA Natural Resource Conservation Service classifies cropland as HEL if the potential of a soil to erode, considering the physical and chemical properties of the soil and climatic conditions where it is located, is eight times or more the rate at which the soil can sustain productivity (https://prod. nrcs.usda.gov/Internet/FSE_DOCUMENTS/nrcs143_007707.pdf, accessed September 2017).

not universal and the conservation plans may not include NT. For Iowa specifically, field surveys suggest that some HEL, especially of higher productivity, is not farmed using NT as often as the non-HEL (Schilling et al. 2007; Tomer et al. 2008).

Large and growing body of agronomic literature is warning against the perception that stover can be collected without much productivity and environmental impact. Leaving the corn stover on the fields not only reduces soil erosion, but also maintains soil organic matter, thus contributing to overall soil sustainability for agricultural production (Blanco-Canqui and Lal 2007; Wilhelm et al. 2004, 2007; Karlen et al. 2011) and water quality (Cruse and Herndl 2009; Thomas et al. 2011; Demissie et al. 2012). Most technical analyses of the stover production potential incorporate soil preservation restrictions (Graham et al. 2007), yet the impact of the restrictions on the economically viable land use and stover collection remains understudied. We fill this gap by comparing and contrasting the economically profitable land use choices under unrestricted stover collection versus the case when stover collection is not allowed on a lower soil quality land.

In the following section we present our data and methods. After describing the economic model, we evaluate the impacts of environmental restrictions under alternative stover prices. We summarize the results and outline the directions for future research in the last section.

2 Methods

2.1 Data

Farmers' choices are simulated for Iowa cropland that was in production in 2009. The GIS representation of the cropland comes from the U.S. Department of Agriculture National Agricultural Statistical Service GIS-based remote-sensing cropland data layer (CDL) (Johnson and Mueller 2010). Soil productivity is measured by the Corn Suitability Rating (*CSR*), an index from 0 to 100 with the higher *CSR* values corresponding to the higher land's productivity in crop production. For each CDL grid unit, the *CSR* value and HEL designation come from the Iowa Soil Properties and Interpretations Database (ISPAID) GIS soil data layer (ISU 2004) (Secchi et al. 2009, 2011b). The resulting data covers approximately 96% of the state's 2009 crop land (Kurkalova and Carter 2017). The distribution of Iowa land by *CSR* and HEL is shown in Fig. 1.

2.2 Farmers' Choices

Our model simulates two choices: crop-tillage rotation and stover collection (Fig. 2). We assume that farmers choose between the two crop rotations, CC and CS. Each year of rotation they choose among three tillage systems: conventional tillage (VT),

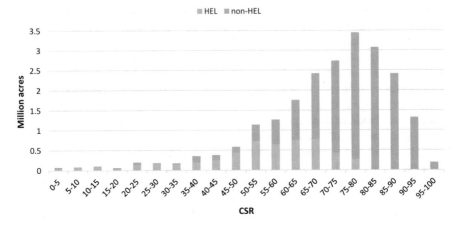

Fig. 1 Distribution of Iowa land by land quality and HEL status

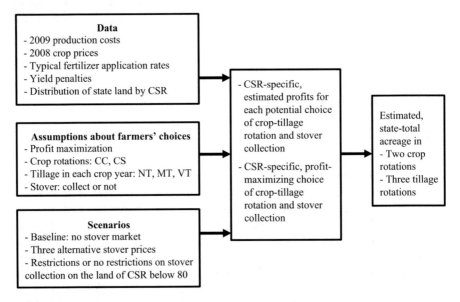

Fig. 2 Overview of the study's modeling approach

mulch till (MT), and NT. Any tillage system that leaves less than 30% residue on soil surface after planting is VT (CTIC 2017). All tillage practices that are not VT are referred to as conservation tillage (CT). Our model distinguishes between two versions of CT, MT and NT. MT disturbs the entire soil surface and involves up to three tillage passes from the harvest of previous crop to the planting of the current crop. NT is an umbrella term used for the tillage systems such as strip-till, vertical tillage, and fluffing harrows, which disturb only the minimal amount of soil (CTIC 2017). Farming operations assumed under specific tillage systems and cropping sequences are based on the typical practices documented by the Iowa State University's Extension (Duffy and Smith 2009). In total, we allow for 9 crop-tillage rotations for CS (both corn and soybean could use one of the three tillage systems). For CC we consider only 6 different crop-tillage rotations, because the order of tillage choices within this rotation does not matter: e.g., we treat corn MT followed by corn NT as equivalent to corn NT followed by corn MT. In contrast, previous studies that explicitly considered crop rotations have restricted tillage to be the same within a rotation, thus allowing for only 3 different crop-tillage rotations for CS and 3 – for CC (Kurkalova et al. 2010; Elobeid et al. 2013).

For each crop-tillage rotation, we model the farmer's choice on whether to collect or not corn stover in the corn production years. Total available stover is equal the corn grain mass produced, but for technological reasons only a fraction of stover is collectable (Graham et al. 2007; USDOE 2016). Following Graham et al. (2007) findings for Iowa, we assume a rate of 67.7% of collection of the total available stover.

The model assumes that production exhibits constant returns to land of any given quality, and that farmers make their choices to maximize 2-year average expected net returns. We treat production input, crop, and stover prices as exogenous; estimate CSR-specific expected net returns under the alternative farmers' choices; and identify the combination of crop-tillage rotation and stover collection choices that maximizes net returns. The crop-tillage rotation component of the model closely follows Kurkalova and Carter (2017), and the stover collection component – that of Kurkalova et al. (2010).

2.3   Model

Expected net returns are the sum of those from (traditional) crop production and those from stove collection.

Expected net returns from crop production are the difference between revenues, which are the product of expected crop price and expected yield, and costs. The 2009 expected prices are assumed equal the prices received at the time of previous year harvest, October 2008 (USDA/NASS 2010): 4.48 $/bu. for corn and 10.4 $/bu. for soybeans (Johanns 2017). Crop yield is the maximum potential crop yield adjusted for previous crop and/or tillage system. Following ISPAID, we model the maximum potential yields in 2009 as $80 + 1.6 \ CSR$ for corn and $23.2 + 0.464 \ CSR$ for soybeans.

Using NT or MT often lowers grain yields relative to VT. We use the data based on multiple long-term agronomic studies in Iowa (Yin and Al-Kaisi 2004; Al-Kaisi et al. 2005, 2015, 2016) to estimate the NT and MT yield penalties. Corn after soybeans yields are equal the maximum potential yield, for both VT and MT, and soybeans after corn yields are equal the maximum potential yield under VT. The remaining six crop-tillage combinations result in yield penalties. Mean corn yield penalties are: 92% of the maximum potential yield for corn NT after soybeans NT, and 81%, 87%, and 90% of the maximum potential yield for corn NT, MT, and VT, respectively, after corn. Mean soybean yield penalties are 94% and 95% of the maximum potential yields under NT and MT, respectively. To account for the potential variability of yield penalty across state, we consider four possible values for each yield penalty, equally spaced between the mean minus two standard deviations and the mean plus one standard deviation. We simulate optimal farmers' choices for each of the possible combination of yield penalties, and then summarize the results as the average over 4096 simulations ($4^6 = 4096$).

The costs of production, by crop, previous crop and tillage are based on 2009 typical production budgets developed by Duffy and Smith (2009), from which we separate the yield-dependent components, to account for the effect of the crop yields varying with innate soil productivity (*CSR*). We maintain most of Duffy and Smith (2009)'s input use and input price values except nitrogen fertilizer application rates and nitrogen and phosphate fertilizer prices. We estimate typical profit-maximizing nitrogen fertilizer rates based on the prices for nitrogen and corn using Corn Nitrogen Rate Calculator (ISU 2004) as 199 lb./ac for corn. While these rates are higher than those reported in Duffy and Smith (2009), we surmise that the actual application rates could be even higher: literature agrees that farmers often overuse fertilizer relative to agronomically recommended rates to avoid potential loss in yield associated with uncertainty in weather and soil nutrients levels (Sheriff 2005). We replace nitrogen and phosphate fertilizer prices of Duffy and Smith (2009), which are significantly higher than the year averages, with the ones estimated by applying agricultural producer price index (https://www.ers.usda.gov/data-products/fertilizer-use-and-price.aspx, Accessed September 2017) to the corresponding 2009 prices (Edwards et al. 2009). The resulting prices are equal 0.35 $/lb. for both nitrogen and phosphate.

Expected net returns from stover collection are the difference between stover collection revenues and costs. The revenues are the product of expected stover price, amount of available stover, and the proportion of stover removed. As explained earlier, we set the proportion of stover removal to 67.7%, where the total available stover is in one-to-one ratio to the grain weigh. To covert the corn production estimates reported in bushels to the stover production estimates reported in metric units, we assume that a bushel of corn has the dry mass of 21.5 kg (56 lb. at 15.5% moisture) (Graham et al. 2007).

Following common approach (Edwards 2011; Dumortier 2016; Chen and Li 2016; USDOE 2016), the cost of corn stover removal includes the cost of chopping and raking corn stalks, baling the stover, and replacing lost crop nutrients. In estimating these costs, we follow the approach outlined in Kurkalova et al. (2010). Our estimates of the costs are based on Edwards (2011), adjusted to reflect the yields that vary by *CSR* and crop-tillage rotation, and the 2009 production input prices.

2.4   Simulation Scenarios

We evaluate a total of seven scenarios: one baseline corresponding to the 2009 economic conditions and assuming no stover market, and six scenarios corresponding
to three alternative stover prices and two restrictions on stover collection.

Following previous research, which reported a wide range of corn stover prices
(Archer et al. 2014; Chen and Li 2016; USDOE 2016), we consider three potential
stover prices: $50, $75, and $100 per ton. Each of the potential stover prices is
considered with and without stover collection restrictions.

The levels of stover removal that do not have long-term soil quality effects could
vary by numerous factors including soil type, cropping history, and tillage system
(Wilhelm et al. 2004; USDOE 2016). Most previous stover assessments assume
no stover collection on soils of high erodibility and/or where harvesting stover is
likely to impede sustainable crop production. We follow Graham et al. (2007), who
argues that the no-stover-collection restriction is commonly needed on the lowest
quality soils, and allow for stover collection only on the land of *CSR* 80 and above.
The threshold of 80 was chosen based on estimated baseline: the land with *CSR*
of 80 and above makes up approximately 36% of Iowa land, but being of higher
productivity, produces approximately 40% of all Iowa corn stover, which is in line
with Graham et al. (2007) finding that roughly 38% of Iowa-produced stover could
be collected safely. Additionally, the threshold of *CSR* 80 also implies that virtually
no stover collection would be allowed on the HEL (Fig. 1). To evaluate how our
results are sensitive to the *CSR* threshold choice, we considered two additional
sets of scenarios, with the thresholds at *CSR* 78 and above, and *CSR* 82 and
above. Such thresholds result in 46% and 33% of the total Iowa corn production,
respectively.

3   Results and Discussion

With the focus of this study on the land-use changes of stover collection, the key
outputs of the model are the simulated average crop and tillage rotations. Predicted
crop rotations are summarized in Fig. 3. We summarize tillage rotations by focusing
on three categories: continuous VT (CVT), continuous CT (CCT), and rotational
CT, which refers to the system when VT was practiced in one of the years, with MT
or NT in the other year of the rotation (Fig. 4).

The baseline simulated is reasonably close to the observed 2009 crop rotation data. According to the CDL data (https://nassgeodata.gmu.edu/CropScape/,
accessed September 2017), the share of cropland in CC rotation was 18.9%, which
is close to our estimate of 18.6%.

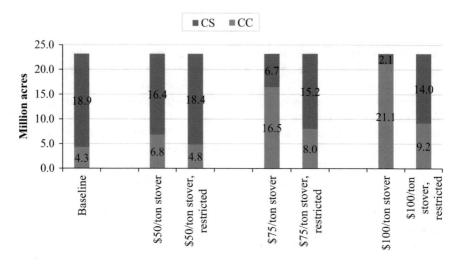

Fig. 3 Crop rotation effect of stover collection. Note: Restricted scenarios assume no stover collection at *CSR* below 80

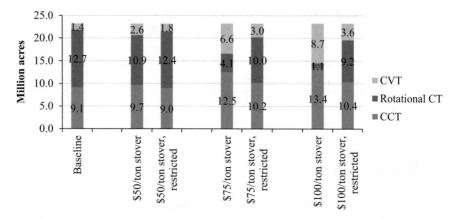

Fig. 4 Tillage rotation effect of stover collection. Notes: Restricted scenarios assume no stover collection at *CSR* below 80. CVT denotes continuous VT, i.e., VT practice in both years of rotation; CCT denotes continuous conservation tillage, i.e., MT or NT practiced in both years of rotation; and Rotational CT means that VT was practiced in one of the years, with MT or NT in the other year of the rotation

We estimate that without stover markets 39% of land is in CCT, 6% is in CVT, and the rest in rotational CT (Fig. 3). The tillage rotation data for comparison with our baseline are not readily available, because such data are rarely collected. A national survey of famers growing corn, soybeans and wheat in the U.S. in 2009 and 2010 found that out of 622 farmers surveyed in the Corn Belt, which includes Illinois, Indiana, Iowa, Missouri, and Ohio, some 55% were in CCT and 14% were in CVT (Andrews et al. 2013). Our estimates imply a single-year CT rate of 66.5%

(39% of CCT plus a half of the 55% of the rotational CT). This is close to the 2012 Census of Agriculture estimates, according to which 67% of Iowa cropland uses CT (https://www.agcensus.usda.gov/, accessed September 2017).

Overall, the impact of unrestricted stover collection on land use is minimal at the lowest stover price we consider, $50/ton, but is noticeable at $75/ton. These findings are consistent with the data on limited sales of corn stover bales at hay auctions in Iowa reported by Edwards (2011), who notes that 0.6-ton bales of corn stover have been sold at the prices from $30 to as high as $45 per bale, i.e., at a price range from $50 to $75 per ton.

The unrestricted stover collection predictably increases the share of land in continuous corn from 19% in the baseline to the estimated 29%, 71%, and 91% under $50, $75, and $100 per ton stover prices, respectively (Fig. 3). Overall, the impact of the stover collection restriction on profit-maximizing crop rotations is remarkable: the prohibition of stover collection on the land below *CSR* 80 affects 9%, 36%, and 52% of the land under $50, $75, and $100 per ton stover prices, respectively.

In the baseline, CC is more prevalent at the higher quality land: only 18.6% of land is in CC at the *CSR* below 80, as opposed to 63% at the *CSR* 80 and above. The comparison between the baseline and the restricted and the unrestricted stover collection scenarios shows that a higher share of the increase in CC comes from the lower quality land. For example, the $75/ton stover price increases CC on some 3.7 million ac of the higher quality land (4.3 vs 8.0 million ac, Fig. 3), and on some 8.5 million ac of the lower quality land (8.0 vs. 16.5 million ac).

Similar to the crop rotations, tillage rotations differ between the higher and the lower quality land. In the baseline, 24% and 1% were in CVT and CCT on the lower quality land, as opposed to 15% and 5% in CVT and CCT on the higher quality land, respectively. Overall, we find that the possibility of selling stover increases CVT and, to a smaller extent, CCT, and as with crop rotations, most of changes in the unrestricted case come from the lower quality land. For example, the $75/ton stover price increases CVT on some 1.6 million ac of the higher quality land (3.0 vs 1.4 million ac, Fig. 4) and on some 3.6 million ac of the lower quality land (3.0 vs. 6.6 million ac). Under the same price, the CCT increases on some 1.1 million ac of the higher quality land (9.1 vs 10.2 million ac) and on some 2.3 million ac of the lower quality land (12.5 vs 10.2 ac). The increase in CVT attributable to the switch from CS rotations to CC has been discussed in the literature as attributable to heavier and sturdier residue that corn has when compared to soybeans, and the subsequent need to more intensive tillage to prepare soils for planting next year crop – to prevent a significant drop in yields (Secchi et al. 2011a). The increase in CCT that we find suggests that the CCT yield loss could be outweigh by the economic benefit of higher net return resulting from growing more corn and lower CT input costs such machinery costs.

Sensitivity analysis reveals that qualitative results are not sensitive to the value of *CSR* below which stover collection is prohibited (Table 1). As discussed above, relatively little land use changes occur at the stover price of $50/ton. In fact, at this stover price, tillage rotations do not change on the higher quality land for all the threshold ranges considered.

Table 1 Crop and tillage rotations under alternative stover collection restrictions

	Crop rotation	Tillage rotation		
Scenario considered	CC (%)	CCT (%)	Rotational CT (%)	CVT (%)
Baseline	19	39	55	6
$50/ton stover	29	42	47	11
$50/ton stover, restricted at CSR below 78	22	39	53	8
$50/ton stover, restricted at CSR below 80	21	39	54	8
$50/ton stover, restricted at CSR below 82	20	39	54	7
$75/ton stover	71	54	18	28
$75/ton stover, restricted at CSR below 78	38	45	41	14
$75/ton stover, restricted at CSR below 80	35	44	43	13
$75/ton stover, restricted at CSR below 82	31	43	46	11
$100/ton stover	91	58	5	37
$100/ton stover, restricted at CSR below 78	43	46	37	17
$100/ton stover, restricted at CSR below 80	39	45	40	16
$100/ton stover, restricted at CSR below 82	35	44	43	14

4 Concluding Comments

Our study complements the growing literature on spatially-explicit, economic rather than only technical, regional assessments of corn stover production in the U.S., such as those for Minnesota (Archer and Johnson 2012), Michigan (Egbendewe-Mondzozo et al. 2011), Indiana (Sesmero and Gramig 2013), and Illinois (Chen and Li 2016). We extend previous assessments for Iowa (Kurkalova et al. 2010; Secchi et al. 2011b; Archer et al. 2014) by advancing the understanding of the interconnection between economically-viable stover collection, crop-tillage rotation, and soil protection restrictions imposed on lower quality land.

We find that if farmers are allowed to participate freely in the stover market, profit maximization is likely to shift cropping patterns towards continuous corn. The effect is accompanied by significant changes in tillage rotations. A notable increase in continuous conventional tillage paired with stover collection is likely to negatively affect the environmental outcomes of crop production such as soil carbon sequestration, soil erosion, nutrient runoff, and in consequence, water quality. We find a smaller potential increase in the use of continuous conservation tillage, but the environmental implications of this change are harder to predict. An increase in conservation tillage is generally associated with improved environmental outcomes, but not in the case when less intensive tillage is paired with stover removal. Investigating the environmental effects of the land use changes presented in this study constitutes a fascinating topic for future research.

We find that imposing restrictions on stover collection on lower quality land alters the land use decisions significantly. From the economic point of view, that means that the restrictions impose a cost on the farmers working lower quality land.

Subsequent analyses could quantify the magnitudes of the opportunity costs, i.e., the potential profits lost for this portion of land managers due to these restrictions. Subsequent analyses could investigate how these costs change with alternative forms of restrictions.

Another possible extension of the current research is to extend the modeling to account for farmers' willingness to participate in the corn stover markets. Recent surveys (Tyndall et al. 2011) indicate that farmers' intent to participate in the stover market may be limited until stover market infrastructure develops and more information on the environmental impacts of stover collection becomes available.

References

Al-Kaisi MM, Yin X, Licht MA (2005) Soil carbon and nitrogen changes as influenced by tillage and cropping systems in some Iowa soils. Agric Ecosyst Environ 105(4):635–647. https://doi.org/10.1016/j.agee.2004.08.002

Al-Kaisi MM, Archontoulis SV, Kwaw-Mensah D, Miguez F (2015) Tillage and crop rotation effects on corn agronomic response and economic return at seven Iowa locations. Agron J 107(4):1411–1424. https://doi.org/10.2134/agronj14.0470

Al-Kaisi MM, Archontoulis SV, Kwaw-Mensah D (2016) Soybean spatiotemporal yield and economic variability as affected by tillage and crop rotation. Agron J 108(3):1267–1280. https://doi.org/10.2134/agronj2015.0363

Andrews AC, Clawson RA, Gramig BM, Raymond L (2013) Why do farmers adopt conservation tillage? An experimental investigation of framing effects. J Soil Water Conserv 68(6):501–511. https://doi.org/10.2489/jswc.68.6.501

Archer DW, Johnson JMF (2012) Evaluating local crop residue biomass supply: economic and environmental impacts. Bioenergy Res 5:699–712. https://doi.org/10.1007/s12155-012-9178-2

Archer DW, Karlen DL, Liebig MA (2014) Crop residue harvest economics: an Iowa and North Dakota case study. Bioenergy Res 7:568–575. https://doi.org/10.1007/s12155-014-9428-6

Blanco-Canqui H, Lal R (2007) Soil and crop response to harvesting corn residues for biofuel production. Geoderma 141:355–362. https://doi.org/10.1016/j.geoderma.2007.06.012

Chen X, Li L (2016) Supply of cellulosic biomass in Illinois and implications for the Conservation Reserve Program. Glob Change Biol Bioenergy 8(1):25–34

Claassen R, Breneman V, Bucholtz S, Cattaneo A, Johansson R, Morehart M (2014). Environmental Compliance in U.S. agricultural policy: past performance and future potential. USDA Economic Research Service. Agricultural Economic Report Number 832. Washington, DC

Conservation Technology Information Center (CTIC) (2017) Tillage type definitions. West Lafayette. http://ctic.paqinteractive.com/media/pdf/TillageDefinitions.pdf. Accessed Sept 2017

Cruse RM, Herndl CG (2009) Balancing corn stover harvest for biofuels with soil and water conservation. J Soil Water Conserv 64(4):286–291. https://doi.org/10.2489/jswc.64.4.286

Demissie Y, Yan E, Wu M (2012) Assessing regional hydrology and water quality implications of large-scale biofuel feedstock production in the Upper Mississippi River Basin. Environ Sci Technol 46:9174–9182. https://doi.org/10.1021/es300769k

Dodder RC, Kaplan PO, Elobeid A, Tokgoz S, Secchi S, Kurkalova LA (2015) Impact of energy prices and cellulosic biomass supply on agriculture, energy, and the environment: an integrated modeling approach. Energy Econ 51:77–87. https://doi.org/10.1016/j.eneco.2015.06.008

Duffy M, Smith D (2009) Estimated costs of crop production in Iowa – 2009. Iowa State University Extension and Outreach, Ag Decision Maker File A1-20, Ames. Revised December 2008

Dumortier J (2016) Impact of agronomic uncertainty in biomass production and endogenous commodity prices on cellulosic biofuel feedstock composition. Glob Change Biol Bioenergy 8(1):35–50. https://doi.org/10.1111/gcbb.12238

Edwards W (2011) Estimating a value for corn stover. Iowa State University Extension and Outreach, Ag Decision Maker FM-1867, File A1-70, Ames. Revised December 2011

Edwards W, Smith D, Johanns A (2009) 2009 Iowa farm custom rate survey. Iowa State University Extension and Outreach, Ag Decision Maker FM-1698, File A3-10, Ames. Revised March 2009

Egbendewe-Mondzozo A, Swinton SM, Izaurralde CR, Manowitz DH, Zhang X (2011) Biomass supply from alternative cellulosic crops and crop residues: a spatially explicit bioeconomic modeling approach. Biomass Bioenergy 35:4636–4647. https://doi.org/10.1016/j.biombioe.2011.09.010

Elobeid A, Tokgoz S, Dodder R, Johnson T, Kaplan O, Kurkalova L, Secchi S (2013) Integration of agricultural and energy system models for biofuel assessment. Environ Model Softw 48:1–16. https://doi.org/10.1016/j.envsoft.2013.05.007

Graham RL, Nelson R, Sheehan J, Perlack RD, Wright LL (2007) Current and potential U.S. corn stover supplies. Agron J 99(1):11

Iowa State University (ISU) (2004) Soil survey and digital soil data: ISPAID version 7.1. http://www.extension.iastate.edu/soils/ispaid. Accessed Sept 2017

Johanns AM (2017) Iowa cash corn and soybean prices. Iowa State University Extension and Outreach, Ag Decision Maker File A2-11, Ames. Revised September 2017. https://www.extension.iastate.edu/agdm/crops/pdf/a2-11.pdf. Accessed Sept 2017

Johnson DM, Mueller R (2010) The 2009 cropland data layer. Photogramm Eng Remote Sensing 76(11):1201–1205

Karlen DL, Birell SJ, Hess JR (2011) A five-year assessment of corn stover harvest in central Iowa, USA. Soil Tillage Res 115–116:47–55. https://doi.org/10.1016/j.still.2011.06.006

Kurkalova LA, Carter L (2017) Mobile technology and sustainable production: using the resource-based view to assess the value of a Green IT artifact. Decis Support Syst 96:83–91. https://doi.org/10.1016/j.dss.2017.02.006

Kurkalova LA, Tran DQ (2017) Is the use of no-till continuous or rotational? Quantifying tillage dynamics from time-ordered spatially aggregated data. J Soil Water Conserv 72(2):131–138. https://doi.org/10.2489/jswc.72.2.131

Kurkalova LA, Secchi S, Gassman PW (2010) Corn stover harvesting: potential supply and water quality implications. In: Khanna M, Zilberman D, Scheffran J (eds) Handbook of bioenergy economics and policy. Springer, New York, pp 307–326

Plourde JD, Pijanowski BC, Pekin BK (2013) Evidence for increased monoculture cropping in the central United States. Agric Ecosyst Environ 165:50–59. https://doi.org/10.1016/j.agee.2012.11.011

Sarica K, Tyner W (2013) Analysis of US renewable fuels policies using a modified MARKAL model. Renew Energy 50:701–709. https://doi.org/10.1016/j.renene.2012.08.034

Schilling KE, Tomer MD, Gassman PW, Kling CL, Isenhart TM, Moorman TB, Simpkins WW, Wolter CF (2007) A tale of three watersheds: nonpoint source pollution and conservation practices across Iowa. Choices 22(2):87–95

Secchi S, Gassman PW, Williams JR, Babcock BA (2009) Corn-based ethanol production and environmental quality: a case of Iowa and the Conservation Reserve Program. Environ Manag 44:732–744

Secchi S, Gassman PW, Jha M, Kurkalova LA, Kling CL (2011a) Potential water quality changes due to corn expansion in the Upper Mississippi River basin. Ecol Appl 21(4):1068–1084

Secchi S, Kurkalova LA, Gassman PW, Hart C (2011b) Land use change in a biofuels hotspot: the case of Iowa, USA. Biomass Bioenergy 35(6):2391–2400. https://doi.org/10.1016/j.biombioe.2010.08.047

Sesmero JP, Gramig BM (2013) Farmers' supply response, price of corn residue, and its economic viability as an energy feedstock. Bioenergy Res 6:797–807. https://doi.org/10.1007/s12155-013-9300-0

Sheriff G (2005) Efficient waste? Why farmers over-apply nutrients and the implications for policy design. Rev Agric Econ 27(4):542–557. https://doi.org/10.1111/j.1467-9353.2005.00263.x

Stern AJ, Doraiswamy PC, Akhmedov B (2008) Crop rotation changes in Iowa due to ethanol production. In: Geoscience and remote sensing symposium. IGARSS 2008, IEEE International, vol V5, pp V200–V203

Thomas MA, Engel BA, Chaubey I (2011) Multiple corn stover removal rates for cellulosic biofuels and long-term water quality impacts. J Soil Water Conserv 66(6):431–444. https://doi.org/10.2489/jswc.66.6.431

Tomer MD, Moorman TB, James DE, Hadish G, Rossi CG (2008) Assessment of the Iowa River's South Fork watershed: Part 2. Conservation practices. J Soil Water Conserv 63(6):371–379. https://doi.org/10.2489/jswc.63.6.371

Tyndall JC, Berg EJ, Colletti JP (2011) Corn stover as a biofuel feedstock in Iowa's bio-economy: an Iowa farmer survey. Biomass Bioenergy 35:1485–1495. https://doi.org/10.1016/j.biombioe.2010.08.049

U.S. Department of Agriculture, Economic Research Service (USDA/NASS) (2010) Field crops: usual planting and harvesting dates, Agricultural handbook number 628. NASS, U.S. Department of Agriculture, Washington, DC

U.S. Department of Energy (USDOE) (2016) 2016 billion-ton report: advancing domestic resources for a thriving bioeconomy, Volume 1: Economic availability of feedstocks. In: Langholtz MH, Stokes BJ, Eaton LM (Leads), ORNL/TM-2016/160. Oak Ridge National Laboratory, Oak Ridge, 448p. doi: https://doi.org/10.2172/1271651

Wilhelm WW, Johnson JMF, Hatfield JL, Voorhees WB, Linden DR (2004) Crop and soil productivity response to corn residue removal: a literature review. Agron J 96(1):1–17

Wilhelm WW, Johnson JMF, Karlen DL, Lightle DT (2007) Corn stover to sustain soil organic carbon further constrains biomass supply. Agron J 99:1665–1667. https://doi.org/10.2134/agronj2007.0150

Yin X, Al-Kaisi MM (2004) Periodic response of soybean yields and economic returns to long-term no-tillage. Agron J 96(3):723–733. https://doi.org/10.2134/agronj2004.0723

Spatial-Temporal Change of Agricultural Biomass and Carbon Capture Capability in the Mid-South of Hebei Province

Yucui Zhang, Qiaoli Hu, Dengpan Xiao, Xingran Liu, and Yanjun Shen

Abstract As an essential part of terrestrial ecosystems, farmland plays a critical role in the carbon cycle. The spatial and temporal characterization of farmland biomass and carbon sequestration capacity is important to understand the carbon cycle of a farmland system. The study area is located in mid-south Hebei Province (MSHP), which is a food production region in North China. Based on land-use data (1980, 1990, 2000 and 2008) and food production data (1984–2008), agricultural biological productivity and carbon capture capacity were estimated. In addition, the spatial-temporal characteristics and related influencing factors were analyzed. Regionwide, aboveground biomass increased from 600 g $C \cdot m^{-2} \cdot a^{-1}$ (1985) to 1200 g $C \cdot m^{-2} \cdot a^{-1}$ (2008) with an increase-decrease-increase pattern during the same period. Spatially, it increased in the piedmont plains and declined in the western mountains and piedmont plains. The carbon capture capacity of cropland in the piedmont area increased from 700 g $C \cdot m^{-2} \cdot a^{-1}$ to 1000 g $C \cdot m^{-2} \cdot a^{-1}$, and it declined in the low plain area. Mountainous and coastal areas had the lowest capability of agricultural carbon capture. Although farmland is a dynamic carbon pool overall, its carbon sequestration capacity is likely to be enhanced with proper farming practices.

Keywords Grain · Carbon sequestration · Biomass · Harvest index · Carbon capture capability · Net primary productivity · LUCC

Y. Zhang · X. Liu · Y. Shen (✉)
Key Laboratory of Agricultural Water Resources & Hebei Key Laboratory of Agricultural Water-Saving, Center for Agricultural Resources Research, Institute of Genetics and Developmental Biology, Chinese Academy of Sciences, Shijiazhuang, China
e-mail: yczhang@sjziam.ac.cn; yjshen@sjziam.ac.cn

Q. Hu
Zhongke Haihui Technology Co., Ltd, Beijing, China

D. Xiao
Institute of Geographical Sciences, Hebei Academy of Sciences, Shijiazhuang, China

© Springer International Publishing AG, part of Springer Nature 2018
R. Li, A. Monti (eds.), *Land Allocation for Biomass Crops*,
https://doi.org/10.1007/978-3-319-74536-7_9

1 Introduction

Farmland on the North China Plain (NCP) is mainly concentrated in the mid-south Hebei Province (MSHP). In China in 2008, approximately 5.05% of grain was produced in the MSHP though it only contains 3.28% of the national cultivated land. Under the background of global change, MSHP, as a high-density population and food-producing area in China, is not only responsible for human food security but also participates in carbon sequestration and carbon balance. Recently, energy conservation and emission reduction have gradually been developing trends. Widely distributed farmland with high productivity is very important for seasonal carbon sequestration, which is significant for the density adjustment of CO_2. By improving agricultural measurements, carbon fixed by plant photosynthesis can be stored in the soil permanently (Potter et al. 1998; Trumbore et al. 1996; Lal et al. 1999). Therefore, it is valuable and practical to conduct research on farmland biological productivity and carbon sequestration.

As early as the beginning of the twentieth century, the Danish plant physiologist Boysen (1932) opened a new era of study on biological productivity using experimental methods on photosynthesis. Subsequently, a number of scientists from Britain and Japan and other countries have conducted further studies regarding biological productivity and have laid a solid foundation. International Biological Planning (IBP) was established in 1964 under the leadership of C.H. Waddington and the International Union of Scientific Societies (ICSU) (Hagen 1992). Biomass productivity is the material production capacity of individuals and groups to ecosystems, regions and the biosphere. Biomass productivity determines the material circulation and energy flow of the system and is considered to be an important indicator of the health status of a system.

Biomass research is closely integrated with carbon sequestration function under global change. Carbon dioxide and other greenhouse gases are absorbed by the photosynthesis of plants, and biomass forms as plants grow. According to the IPCC (2000) report, global carbon storage capacity has reached 466 billion tons, accounting for approximately 20% of total land carbon storage. The biological productivity of vegetation, especially the biomass and carbon capture of vegetation, is an important parameter or variable that characterizes the growth of vegetation. It not only reflects the growth status of vegetation but also reflects the ability of vegetation to intercept greenhouse gases. Biomass is one of the key variables used to characterize plant activities. It is not only the basis of biological productivity, net primary productivity, and global change research but also has a certain practical significance. Similarly, studies of terrestrial biomass are helpful in understanding and investigating global carbon storage, cycling and balance. However, research on biomass is mainly focused on natural vegetation such as forests and grasslands, and less so on farmland. Therefore, the study of biomass in farmlands has important significance.

With concerns of greenhouse gas emissions, carbon sinks and carbon cycles have become one of the hot spots of current research. Researchers have mainly focused on forest ecosystems and less so on farmland ecosystems (Kotto-Same et al. 1997; Pathak et al. 2011; Smith et al. 2008; Usuga et al. 2010; Svirejeva-Hopkins et al. 2004; Riedel et al. 2008; Yoshimura et al. 2010). A farmland ecosystem is affected by human activities, and has the dual identity of "source" and "sink." Soil carbon emissions, fertilizer and pesticide consumption, and agricultural machinery are the main carbon sources in farmland ecosystems (West and Marland 2002). Straw burning will elevate carbon emissions, but this practice is very scarce in the NCP today. Yu et al. (2009) noted that farmland had changed from a carbon source to a carbon sink in China. Although many researchers believe that farmland has great potential for carbon sequestration, the study of agro-ecosystems is still very complicated due to the complexity of farmland ecosystems and strong human intervention (Schneider 2000; Schneider et al. 2007). Related research on carbon sinks and the carbon cycle is mainly concentrated on a global, national, or other large regional scale. Piao et al. (2001, 2009, 2011) studied carbon sequestration and the carbon budget of land systems in China. Fang et al. (2001) analyzed the carbon storage of forest ecosystems (1949–1998) and land vegetation carbon sequestration (1981–2000) in China.

Models used in carbon cycle research can be divided into three types: empirical models represented by the Miami model (Lieth 1972; Lieth and Whittaker 1975) and the Thornthwaite Memorial model (Lieth and Box 1972), semi-empirical semi-mechanical models represented by the Chikugo model (Uchijima and Seino 1988; Jiang et al. 1999), and mechanism models represented by the CASA (Potter et al. 1993) and CENTURY (Parton et al. 1987; Parton and Rasmussen 1994) models. The CASA model was a light use efficiency model driven by remote sensing data and data of vegetation and soil types, solar radiation, temperature, precipitation and other parameters. The mechanism model compensates for the disadvantages of the empirical and semi-empirical model, and the computational accuracy is improved to some extent. However, they needed climatic data and even some empirical formulas which could lead to poor automation and real-time processing, thus reducing the accuracy of the model (Zhou et al. 1998). In our research, mainly based on grain yields and land use data, a simple model of carbon sequestration was constructed. In this study, data regarding crop yield and land use from 1984 to 2008 were used to estimate farmland biomass and carbon interception in the MSHP. The temporal-spatial distribution characteristics and variations were also analyzed. The impacts of anthropogenic carbon capture on human carbon sequestration capacity and its mechanism were analyzed to provide data support and a research basis for the carbon cycle of a farmland system. This research will provide scientific support for national policy and farmland management.

2   Study Area and Methods

2.1   Study Area

The farmland in this study is mainly concentrated in the MSHP. The study area is between 36°02′48″ to 39°57′02″N and 113°27′21″ to 117°47′49″E, with a total area of 88,721 km². Farmland in this area accounts for over 60% of the total amount of that occurring on the NCP (Fig. 1).

This region is a warm temperate semi-humid and semi-arid continental monsoon climate. It is very cold in the winter but hot and rainy in the summer. The annual sunshine hours are 2500   3000 h. Daily temperatures of >10 °C are about 100–210 days year^{-1}, with an annual accumulated temperature of 1600   4600 °C. Annual average precipitation is approximately 530 mm, mostly occurring from June to September. Winter wheat and summer maize are the most important crops, and other crops such as cotton, soybean, rape, millet, vegetables, and fruits are also planted. The total sown area of crops was approximately 8.71 million ha in 2008. The grain sown area was 6.16 million ha (approximately 70.68%), while cotton was 0.69 million ha (approximately 7.92%) and bean was 249,600 ha (2.86%). The sown area of vegetables was 1.1 million ha, accounting for 12.64% of the total sown area. Woodland is mainly distributed in the western mountains. At the end of 2008, there were 5.29 million ha of forestland and 1.06 million ha of orchards.

2.2   Data Sources

Land Use/Cover Data

Land use/cover data in 4 years (i.e., 1980, 1990, 2000 and 2008) were used, which were available from the Knowledge Innovation Program of the Chinese Academy of Sciences during the Eleventh Five-year Plan. The resolution was 30 m with twenty secondary land classes, which were combined into six primary categories in this study: farmland land, forest, grassland, water body, construction land and unused land (Table 1).

Grain Yields and Other Data

1. Grain yields data

The statistical data for grain yields in the counties were obtained from the Economic Statistical Yearbook of Hebei Province (1985–2009), the Rural Statistical Yearbook of Hebei Province (1995–2009) and Fifty years of New Hebei Province (1999). To analyze the regularity of grain yields and to prevent the volatility of data in individual years, this study used an average of 5 years to represent the grain yields of a certain period.

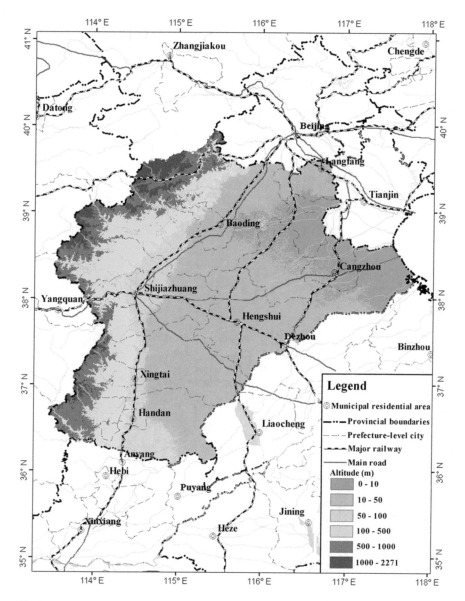

Fig. 1 Location of the study area

Table 1 Classification table of land use / cover

Old land use classification		New land use classification		Old land use classification		New land use classification	
Code	Type	Code	Type	Code	Type	Code	Type
11	Paddy field	1	Farmland	114	Glacier and snow	4	Water area
13	Dry land			115	Intertidal zone		
21	Orchard	2	Forestland	71	Urban land	5	Construction land
31	Forest land			72	Rural residential land		
32	Shrub			61	Independent industrial and mining land		
33	Open forest land			101	Railway land		
41	Grassland	3	Grassland	102	Highway land		
111	Graff	4	Water area	116	Beach land	6	Unutilized land
112	Lake			125	Wetland		
113	Reservoir						

Table 2 Harvest index of different crops

	Wheat	Maize	Oil crops
1985	0.40	0.27	0.43
1990	0.33	0.32	0.41
1995	0.40	0.42	0.40
2000	0.49	0.48	0.44
2005	0.54	0.51	0.45
2008	0.54	0.54	0.48

2. Harvest index

The harvest index (HI) refers to the ratio of grain yield and total aboveground biomass. The harvest index is varies between different crops and different periods. To simplify the calculation of biomass, crops are classified as wheat, maize, oil crops and cotton. According to previous research (Xue 2004; Yuan and Guan 1997; Yuan et al. 1999; Tian et al. 2007), the HIs of different crops are shown in Table 2. As the same or similar research area, the harvest index used in this paper is reasonable.

2.3 Study Methods

The carbon capture model was constructed based on grain yield data, land use data and other relevant agricultural data. Then, the temporal and spatial distribution of the carbon intercept was analyzed. Finally, the influencing factors were analyzed and evaluated.

Aboveground Biomass of Farmland

Biomass refers to the total amount of biomass accumulated per unit area over a certain period, the size of which directly reflects the biomass productivity in a region. In agricultural production, crop biomass usually refers to the total dry matter weight of aboveground straw and grain. Therefore, the aboveground biomass is calculated as follows:

$$B_{i,j} = \frac{Y_{i,j}}{HI_{i,j}} \tag{1}$$

where $B_{i,j}$ represents the aboveground biomass of crop j in the unit area of year i (g m^{-2} a^{-1}); $Y_{i,j}$ is the grain yield of the crop j in year i (g m^{-2} a^{-1}); and $HI_{i,j}$ is the harvest index of crop j in year i.

Carbon Capture Capability of Farmland

Carbon capture capacity is the capacity of crops to assimilate CO_2 in the atmosphere during photosynthesis, usually expressed in terms of carbon equivalents (unit: g C $m^{-2} a^{-1}$). In agricultural systems, the carbon capture capacity of crops is the carbon content of the crop at the end of a growth period, that is, the total carbon content of dry matter. The carbon capture capacity of crops is calculated from the biomass of crops, and the formula is as follows:

$$C_i = \sum_j \left[B_{i,j} \times (r_j + 1) \right] \times \alpha \tag{2}$$

where C_i is the carbon capture capacity of one crop in the unit area during year i; $B_{i,j}$ is the aboveground biomass of the crop j in the unit area in year i; r_j is the average root cap ratio of the crop j; and α is the average carbon content of the crop.

1. Root cap ratio (r)

Root cap ratio refers to the ratio between the underground part of the crop and the aboveground part. Zhang (1999) found that the ratio of root to shoot was approximately 0.15–0.20 during the growth period of winter wheat at the Luancheng Experimental Station. The root cap ratio of mature maize was 0.10–0.15 while sorghum was much higher than that of wheat, maize, and millet, changing from 0.80 at the early growth stage to 0.53 when harvested. This study used the root cap ratio of wheat, maize, millet and sorghum, which was 0.15, 0.10, 0.23 and 0.53, respectively.

2. Carbon-content ratio of crops (α)

Although the chemical compositions of different crops are very different, the carbon intercepted by crops is fixed in the form of photosynthate. In addition, the amount of CO_2 captured in the atmosphere was estimated by biomass. The chemical equation for photosynthesis is as follows:

$$(6CO_2 + 5H_2O) \times n \xrightarrow{photosynthesis} (C_6H_{10}O_5) n + 6nO_2 \tag{3}$$

And simplified to the following:

$$6CO_2 + 5H_2O \xrightarrow{photosynthesis} C_6H_{10}O_5 + 6O_2 \tag{4}$$

From Eq. (4), it can be seen that the CO_2 is fixed to the crop as a form of photosynthate. Carbon content of photosynthates accounted for approximately 44.44%, and the determination of crop interception accounted for 44.44% of the total biomass.

In this research, land use/cover data (1980, 1990, 2000, and 2008) were scaled up to extract the percentage of cultivated land and farmland. Spatial interpolation

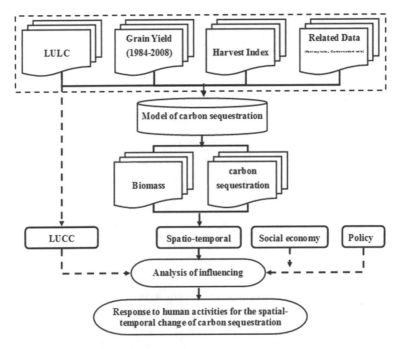

Fig. 2 The scheme and structure of this research (LUCC: Land - Use and Land - Cover Change; LU/LC: land use / land cover)

was used to obtain land use/cover data for another 3 years (1985, 1995, and 2005). Biomass of the crop was obtained from the years of grain production statistics and the intercultural harvest index for different crops. Combined with the ratio of root to shoot and the carbon content of different crops, carbon capture capacity was obtained. Finally, the factors affecting the carbon capture capacity of farmland were analyzed. The detailed research scheme is as follows (Fig. 2).

3   Spatial–Temporal Distribution of Carbon Capture Capability on Farmland

3.1   Biomass of Farmland in the Mid–South Hebei Province

Since the 1980s, the level of agricultural production has been highly improved with the development of a social economy. For example, the input of chemical fertilizer increased from 18.39 g m^{-2} a^{-1} in 1985 to 49.84 g m^{-2} a^{-1} in 2008, and irrigation water has been further guaranteed. Therefore, the grain yield has significantly increased and has nearly doubled since the 1980s in the MSHP (Fig. 3a). However,

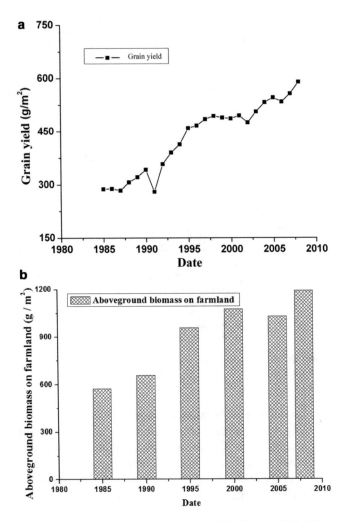

Fig. 3 The annual grain yield (**a**) and biomass (**b**) in the Mid – South of Hebei Province (MSHP)

many farmlands were abandoned as increased yields did not increase income. Together with droughts and floods, the yields in 1991, 2002 and 2006 decreased.

With increasing grain yield, the aboveground biomass on farmland increased from approximately 600 g m^{-2} a^{-1} in 1985 to 1200 g m^{-2} a^{-1} in 2008 (Fig. 3b). The biomass increased quickly although it decreased from 2003–2005 due to the yield. Therefore, the biomass productivity is increasing in the MSHP. The biomass in most areas increased 50–500 g m^{-2} a^{-1} from 1985 to 1990. In the early 1990s, the biomass still increased 50–500 g m^{-2} a^{-1} except in some areas such as Shijiazhuang and Baoding. From the late 1990s to the early 2000s, the biomass began to decrease. Especially from 2000 to 2005, the biomass decreased 50–500 g m^{-2} a^{-1} or more.

However, biomass increased again by 50–500 g m^{-2} a^{-1} from 2005 to 2008. During the period 1985–2008, the biomass in the western mountainous area and the piedmont such as in Shijiazhuang and Baoding decreased, and the biomass on the low plain such as in Heilonggang increased (Fig. 4). This phenomenon appeared mainly because improved agricultural technology and increasing agricultural input enhanced the biomass on the low plain. While agriculture developed well on the piedmont, biomass in some areas decreased as people shifted to plant economic crops with higher benefit. Although the grain yield and biomass increased for more than 20 years, original fixed carbon is released by consuming grains and burning and decomposing straw, which brings carbon and nitrogen cycling back again. Thus, increasing crop biomass accelerated the carbon and nitrogen cycle.

3.2   Spatial-Temporal Distribution of Carbon Capture Capability of Farmland

Biomass has increased rapidly over more than 20 years, and the carbon capture capability has also improved. Farmland on the piedmont had the strongest capability to sequester carbon, following by farmland on the low plain. The carbon capture capability in the mountainous area was the weakest. The carbon was maximally captured by crops on the piedmont of Taihang Mountain at 700–1000 g C m^{-2} a^{-1}, which was 400 g C m^{-2} a^{-1} more than that of crops on the low plain (Fig. 5). The crops in the mountainous and coastal areas captured lower carbon at approximately 150 g C m^{-2} a^{-1}. Overall, the carbon captured by crops in the whole region has increased from 0.021 Pg C to 0.025 Pg C during the past 20 years. The carbon was mostly captured by crops on the piedmont of Taihang Mountain because of the fertile soil and good agricultural production conditions there. The capture capability in the western mountainous area was low because the carbon sequestration of grasslands and forest was not considered in this study. For the low plain, agricultural production is relatively low which leads to low carbon sequestration. The main crop is cotton in the mid–south of the study region, which was not considered in this study, either. Thus, the carbon sequestration of this area also appeared low.

During crop growth, most CO_2 is fixed into plant bodies to form biomass, and very little is fixed into soil carbon pools. To further explore the spatial-temporal distribution of the carbon capture capability of farmland, points on two sample lines were selected from west to east along longitude 115°E and south to north along latitude 38°N. From low to high longitude, the carbon sequestration first increased, and then decreased. The sequestration was highest between 114.75° and 115.75°E. The line from west to east crossed mountainous areas, the piedmont plain and the low plain. Both the total carbon sequestration and variation were greater on the piedmont plain over the last 20 years. Although the carbon sequestration changed less on the low plain, it is increasing. For the mountainous areas, the carbon capture capability was lower and has reduced slightly (Fig. 6 left).

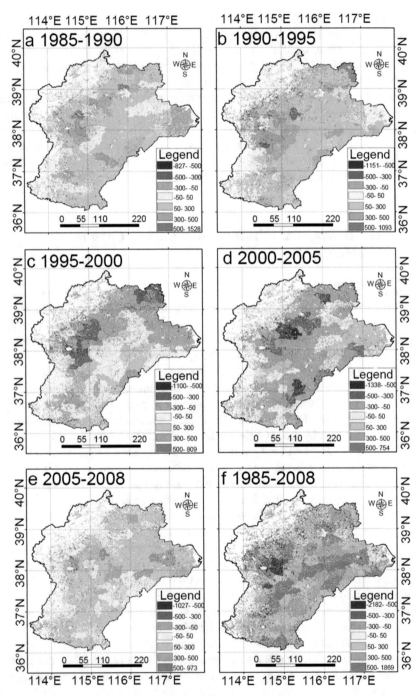

Fig. 4 Spatial-temporal variation of aboveground biomass in the MSHP (unit: g m^{-2} a^{-1})

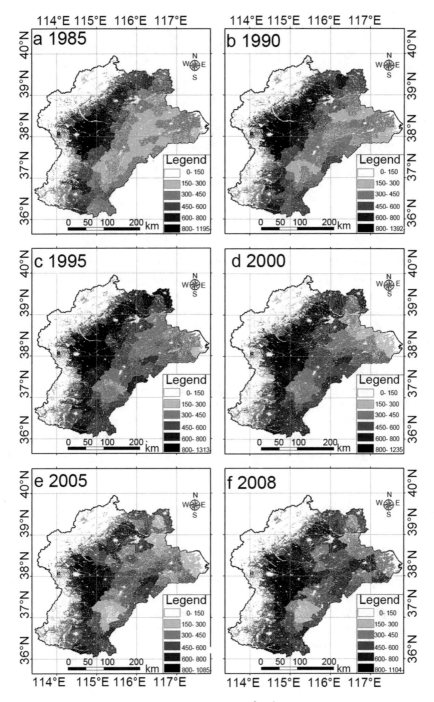

Fig. 5 Carbon sequestration in the MSHP (unit: g C m^{-2} a^{-1})

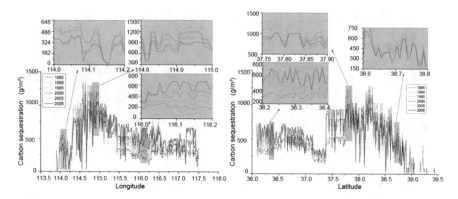

Fig. 6 The samples distribution of carbon sequestration (Left: 38°N; Right: 115°E)

The carbon capture capability showed some regular spatial-temporal changes with latitude, and it also showed different spatial-temporal changes along longitude. There were two peaks of carbon sequestration: one was in Shijiazhuang at approximately 38°N, and the other was in Handan at approximately 36.5°N (Fig. 6 right). By selecting sample points in Handan, Shijiazhuang, and Baoding, carbon sequestration was analyzed with longitude. The carbon capture capability became stronger in Handan, while in Shijiazhuang and Baoding it first increased and then decreased slightly during the 1990s.

Three sample points (38.00°N, 114.50°E; 38.00°N, 114.75°E; and 38.00°N, 116.00°E), which respectively represent the mountainous area, the piedmont plain and the low plain, were selected to analyze carbon capture capability (Fig. 7). The results showed that the carbon capture capability decreased in the mountainous area and on the piedmont plain, but increased on the low plain.

3.3 Comparison to Data from the Literature

To evaluate accuracy, the results were compared to other research results. In this study, carbon sequestration is the carbon content of the crop per unit time. Net primary productivity (NPP) is the remaining fraction of carbon sequestration by photosynthesis per unit time after accounting for energy lost due to cellular respiration and maintenance of plant tissue. Considered from a different perspective, carbon sequestration and NPP both refer to the organic carbon content. Compared to the results estimated by Chen et al. (2011) using the CASA model (Fig. 8), the average carbon sequestration of farmland fluctuated between 370 and 460 g C m^{-2} a^{-1}, which was 120 g C m^{-2} a^{-1} less than the result of Chen et al. (2011). However, the NPP calculated by Chen et al. (2011) was based on mixed pixels, which included crop, grass, trees, etc., while the carbon sequestration of only grain crops was calculated in this study. Therefore, the results were less than those

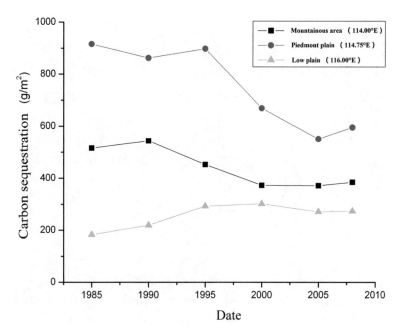

Fig. 7 Samples distribution of three typical (38°E)

estimated by Chen et al. (2011). However, their trends were basically the same. Piao et al. (2001) estimated the annual NPP of most natural vegetation in temperate regions except forests which was between 200 and 400 g C m^{-2} a^{-1} in China. Sun and Zhu (2000) studied the NPP in China in 1992. The NPP estimated using a climatic model was 396.36 g C m^{-2} a^{-1} for an annual double-crop rotation system, while the NPP estimated using the Miami model was 497.89 g C m^{-2} a^{-1}. Wang et al. (2006) estimated the agricultural NPP in China using Crop-C and it increased from 300–500 g C m^{-2} a^{-1} in 1985 to 500–700 g C m^{-2} a^{-1}. Li (2004) estimated the NPP in China using MODIS data and the CASA model, and the NPP in Hebei Province and on the North China Plain in 2001 was 382 g C m^{-2} a^{-1} and 300–550 g C m^{-2} a^{-1}, respectively. Liu et al. (2010) analyzed the NPP in North China using the CASA model and NOAA/AVHRR data, and found that the NPP of farmland in North China in 2007 was 363–376 g C m^{-2} a^{-1}. Although the data and methods might be different in these studies, the results in the region are similar.

4 Factors of Carbon Interception Capacity by Farmland

Changes in biomass and carbon interception capacity varied over time and space, mainly on the piedmont plain, where these were the highest. However, biomass and carbon intercept declined with time on the piedmont plain and mountainous

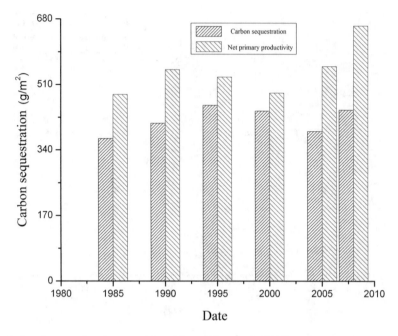

Fig. 8 Result comparison between this article and Chen et al. (2011)

farmland, while the trend was the opposite on the low plain. Therefore, the carbon interception capacity on the piedmont plain and mountainous farmland gradually declined, while it increased on the low plain. To determine the impact factors and trend of carbon interception capacity, the impacts of land use/cover change, agricultural production conditions and other factors were analyzed.

4.1 Response of Farmland Carbon Interception to Land-Use and Land-Cover Change

During the past 30 years, construction on land in urban and rural areas has covered significant areas of high-quality farmland, leading to a continuous decrease in farmland (Liu et al. 2009). Forest land and grassland, mainly affected by economy and policy, have also been reclaimed or ecologically restored (Ge and Dai 2005). Therefore, land use/land cover (LU/LC) change occurred at different levels. The acceleration of urbanization has resulted in many cultivated lands being converted into construction land, especially in Shijiazhuang and other large- and medium-sized cities (Fig. 9) (Xiaoet al. 2006). As an important component of terrestrial ecosystems, changes in LU/LC not only change the properties of surface cover but

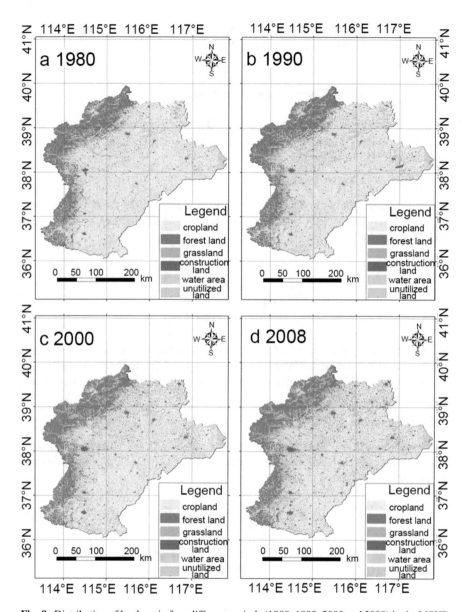

Fig. 9 Distribution of land use in four different periods (1980, 1990, 2000 and 2008) in the MSHP

also exert impacts on the carbon cycle, water cycle, regional climate, ecological environment and biodiversity (Guo et al. 1999; Gao et al. 2007).

Methods of Land Use/Cover Change Detection

To quantitatively estimate the changes in LU/LC on the central and southern plain of Hebei Province, a conversion matrix and dynamic degree of land use were used to quantitatively assess the LUCC.

1. Conversion matrix of land use

By detecting changes in land use at different time series, the direction and structural characteristics of LUCC can be determined. The equation of the conversion matrix was as follows:

$$S_{i,j} = \begin{bmatrix} S_{11} & S_{12} & \cdots & S_{1n} \\ S_{21} & S_{22} & \cdots & S_{2n} \\ \cdots & \cdots & \cdots & \cdots \\ S_{n1} & S_{n2} & S_{n3} & S_{nn} \end{bmatrix} \tag{5}$$

where S represents the area of land use; n represents the number of land use types; i, j represent the types of land use at the beginning and the end of the study period, respectively; and S(i, j) is the area of the land-use type, which is changed from land k to land use of $k + 1$.

2. Dynamic degree of land use

The dynamic degree of land use is used to quantitatively estimate changes in land use, the regional difference in LUCC and the trend of LUCC in the future (Liu et al. 2003; Song et al. 2008) as follows:

$$R = \frac{U_b - U_a}{U_a} \times \frac{1}{T} \times 100\% \tag{6}$$

where U is the number of land use types at the beginning (U_a) and at the end of the study period (U_b), and T is the length of the study period. When T is set as 1 year, the value of R is the annual change rate for a certain land use type in the study area.

Land Use/Cover Change (LU/LC)

The LU/LC of 1980 and 2008 were statistically compared using the land use conversion matrix and the dynamic degree of land use (Table 3). Thearea of farmland significantly decreased, most of which changed into construction land. The change in forestland is not significant, while the rate of grassland growth was as high as 15.5%. The construction land, which mainly originated as farmland, also

Table 3 The conversion matrix and relative change rate of LUCC from 1980 to 2008 (unit: km^2)

1980	2008						
	Farmland	Forest land	Grassland	Construction land	Water area	Unutilized land	Total
Farmland	51592.3	588.4	949.6	4453.6	549.3	817.5	58950.7
Forest land	331.5	11750.6	2120.1	49.1	18.0	52.2	14321.5
Grassland	512.7	1726.2	2641.9	63.8	32.0	54.4	5031.0
Construction land	418.3	9.8	9.3	6019.6	49.6	26.0	6532.6
Water area	504.5	34.4	41.3	131.5	1339.4	410.9	2462.0
Unutilized land	478.9	49.7	49.1	44.3	90.0	712.8	1424.8
Total	53838.2	14159.1	5811.3	10762.0	2078.3	2073.8	88722.5
Amount of change	−5112.6	−162.4	780.3	4229.4	−383.7	648.9	—
Rate of change (%)	−8.7	−1.1	15.5	64.7	−15.6	45.5	—

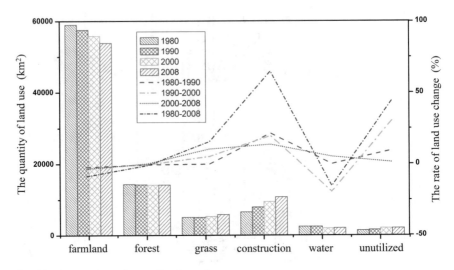

Fig. 10 The quantity and rate of land use change

increased by up to 64.7%. Water area slightly decreased, and most of that change was into farmland or unutilized land (Fig. 10).

Response of Carbon Interception of Farmland to LUCC

In recent years, accelerated urbanization has led to a decline in high-quality farmland (Zhu et al. 2001), which, to some extent, has weakened the carbon interception capacity of farmland. In the past 30 years, the amount of farmland land that has changed into another type of land was significantly higher than that of any other type of land converted into farmland (Fig. 11). From 1980 to 2008, the area of farmland land decreased 5112.6 km^2 (8.7%). Without taking into account other land use types, the interception capacity of carbon in farmland was calculated as 500 g C m^{-2} a^{-1}, resulting in a reduction of farmland carbon interception of approximately 2.6 million tons.

Using re-sampling, the net change of farmland from 1985 to 2008 with a resolution of 5 km × 5 km was obtained (Fig. 12a). As shown in Fig. 12a, farmland on the piedmont plain was significantly reduced and is especially obvious around the Shijiazhuang region and coastal plain. Distribution of carbon density changes for farmland between 1985 and 2008 are shown in Fig. 12b. The amount of carbon capture is significantly reduced in the Shijiazhuang piedmont plain region, while it increased significantly in the fields region with middle and low yield. As shown in Fig. 13, carbon density changes in farmland mainly appeared in the piedmont plain area and around the city. Due to the decrease in farmland, the difference in the carbon capture capability between Shijiazhuang and the other cities has increased.

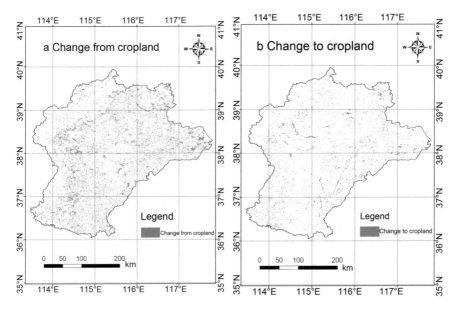

Fig. 11 Changes of cropland area: cropland changed to other land use types (**a**) and other land use types changed to cropland (**b**) (1985–2008, 5 km × 5 km)

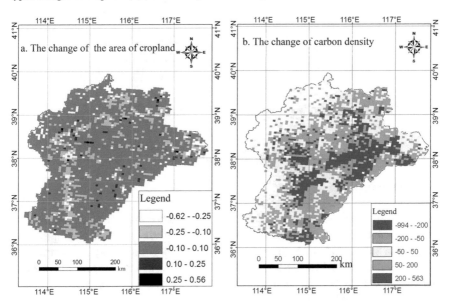

Fig. 12 The change of the area of cropland (**a**) and carbon density (**b**) (1985–2008, 5 km × 5 km)

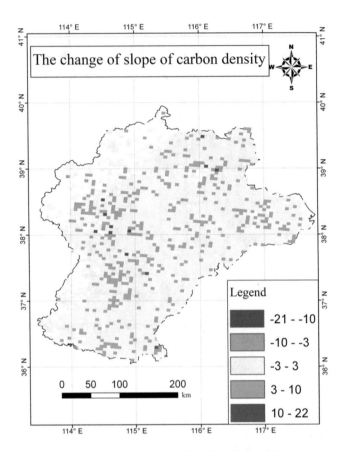

Fig. 13 The change of slope of carbon density (1985–2008, 5 km × 5 km)

4.2 Impacts of Agricultural Management

Fertilization

In addition to the direct impact on farmland, the increase in the fertilizer application rate, the improvement of irrigation, the shift of cultivars and the development of cultivation techniques will affect the carbon interception ability. A comparison of carbon density and fertilizer application is shown in Fig. 14 left. Since 1985, the carbon density and fertilization rate have increased from 300 g m^{-2} a^{-1} and 22 g m^{-2} a^{-1} to 600 g m^{-2} a^{-1} and 50 g m^{-2} a^{-1}, respectively, and have a good correlation (Fig. 14 right). Therefore, the carbon interception capacity of farmland has a close agreement with the amount of fertilizer applied.

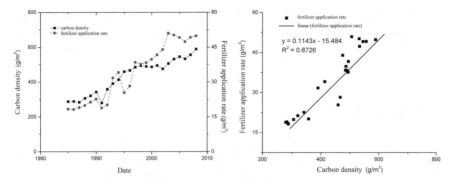

Fig. 14 Comparison of carbon density and fertilizer (left) and the correlation (right)

Irrigation

Since 1984, the rate of irrigation in the central-south region of Hebei Province has rapidly increased (Fig. 15a), with considerable differences in different regions (Fig. 15b). In the Hei-Long-Gang region, where water levels have decreased and the level of salinization has been reduced in recent years, construction of water conservancy facilities and irrigation area have significantly increased. However, in the central Shijiazhuang region and in some counties and cities of Baoding, the irrigation rate has been reduced during the past 20 years (Fig. 15b). The main reason is that many excellent and irrigated farmlands have been developed into construction land, resulting in a decline in the irrigation area. Other regions have an increased irrigation rate mainly due to the construction of water conservancy facilities, which have transformed dry land into farmland.

4.3   Other Factors

Meteorological Conditions

Precipitation can improve carbon interception of farmland by relieving drought. However, too much or too little precipitation will pose a threat to agricultural production. The frequency of droughts and floods in China has increased in recent years and extreme weather events will increase in the future (Liu 2011). In November of 2009, heavy snowfall in Shijiazhuang and other regions damaged agricultural production and a drought during the winter of 2010 also brought a serious threat to food production. There was a significant anticorrelation between the frequency of extreme droughts () and grain yield both in the north (Shanxi) and south (Guangzhou) of China (Fig. 16 left, from Liu 2011). Temperature also plays an important role in the carbon interception capability of farmland. Crop growth stages is often measured by accumulated temperature conditions. Too high or too low temperature in a short period also causes harm to crop growth and agricultural production, ultimately affecting the carbon interception capability of farmland (Fig. 16 right, from Ferris et al. 1998).

Social Economy

With social and economic development, people have changed the farmland into factories and mines for more economic benefit. Also with an increase in population, there is a need for more land for construction to meet housing, road and other infrastructure needs. Since 1980, approximately 4453.6 km^2 of farmland has been converted to construction land. Due to the reduction in farmland, the carbon interception ability of farmland significantly decreased.

Policy

Policies in China, including returning farmland to forest, ecological restoration and land consolidation measures, have caused changes in farmland productivity. During the past 30 years, 588.4 km^2 and 949.6 km^2 of farmland have been returned to forestland or grassland in northern Hebei Province, respectively. In

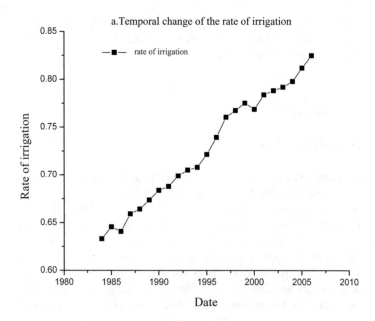

Fig. 15 Temporal (**a**) and spatial (**b**) changes of irrigation rate in the MSHP

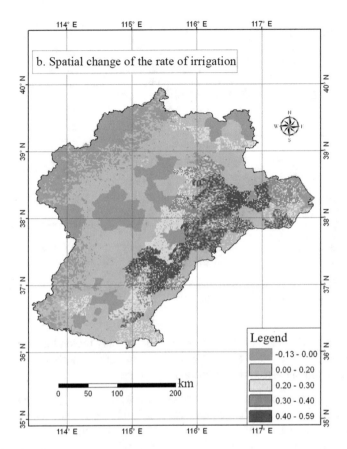

Fig. 15 (continued)

terrestrial ecosystems, forest and grassland contain larger carbon stock. Therefore, the implementation of returning farmland to forest or grass will increase the local capability for carbon interception.

5   Conclusions

With ongoing global climate change, farmland is gradually being changed from a carbon source to a carbon sink. However, the spatiotemporal variation in cropland productivity (biomass) and its potential for carbon capture is key to coping with climate change. Therefore, this study on carbon capture of farmland is of vital significance. In this study, a carbon capture model was used to combine crop yield, land use data and other relevant agricultural data to obtain biomass and carbon capture values of farmland. The temporal-spatial variation in carbon capture capacity was also analyzed. The results indicated that:

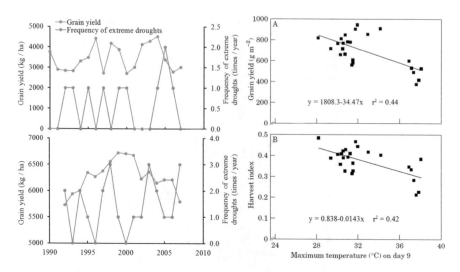

Fig. 16 Relationship of grain yield (or Harvest index) and precipitation (left, from Liu 2011) and temperature (right, from Ferris et al. 1998)

1. Since the 1980s, the biomass of farmland in the study region has increased from 600 g m^{-2} a^{-1} in 1985 to 1200 g m^{-2} a^{-1} in 2008, mainly driven by human activities and the higher use of water and fertilizer.

The carbon capture capacity of farmland increased by nearly 300 g C m^{-2} a^{-1}, and the total amount of carbon capture reached 0.025 Pg C in 2008. Moreover, the piedmont plain area has a higher carbon capture level than that of the low plain area. Our results are similar to other related research results.

2. The increase in the carbon capture capacity of cropland in central and southern Hebei Province has accelerated the process of carbon cycle and carbon sequestration.
3. The total amount of carbon intercepted in farmland is very large, although it is only a dynamic carbon pool. However, it will increase the actual carbon sequestration capacity of farmland if it can be used rationally.

Over the past 30 years, the intensity of land use changes around urban and rural settlements is largely the due to human activities. Land use in Shijiazhuang has changed significantly as a result of human activities. At the same time, human activities have obviously affected land use change on both sides of the road, which has turned farmland into non-agricultural land. As a result, as construction land expands and human activities increase, the ability to capture carbon in nearby areas is gradually weakened. This phenomenon leads to the formation of regular spatial and temporal variation in carbon capture capacity. Under the effect of agricultural production conditions, socioeconomics, policy and other human activities, carbon capture capacity also showed different temporal and spatial characteristics. In summary, human activities exert a direct or indirect impact on farmland, resulting in temporal-spatial variability in cropland carbon capture capacity. Therefore, the government should take some measures to guide beneficial human activities to the sustainable development of the ecological environment.

References

Boysen JP (1932) Die stoffproduktion der pflanzen. G. Fischer, Jena, pp 1–108

Chen F, Shen Y, Li Q, Guo Y, Xu L (2011) Spatio-temporal variation analysis of ecological systems NPP in China in past 30 years. Sci Geogr Sin 31(11):1409–1414

Fang J, Chen A, Peng C, Zhao S, Ci L (2001) Changes in forest biomass carbon storage in China between 1949 and 1998. Science 292:2320–2322

Ferris R, Ellis RH, Wheeler TR, Hadley P (1998) Effect of high temperature stress at anthesis on grain yield and biomass of field-grown crops of wheat. Ann Bot 82(5):631–639

Gao XJ, Zhang DF, Chen ZX, Pal JS, Giorgi F (2007) Land use effects on climate in China as simulated by a regional climate model. Sci China Ser D Earth Sci 50(4):620–628

Ge QS, Dai JH (2005) Farming and forestry land use changes in China and their driving forces from 1900 to 1980. Sci China Ser D Earth Sci 48(10):1747–1757

Guo XD, Chen LD, Fu BJ (1999) Effects of land use/land cover changes on regional ecological environment. Adv Environ Sci 7(6):66–75

Hagen JB (1992) An entangled bank. Rutgers University Press, New Brunswick

IPCC (2000) Land use, land-use change, and forestry. IPCC, Cambridge

Jiang H, Apps MJ, Zhang Y, Peng C, Woodard PM (1999) Modelling the spatial pattern of net primary productivity in Chinese forests. Ecol Model 122(3):275–288. https://doi.org/10.1016/s0304-3800(99)00142-8

Kotto-Same J, Woomer PL, Appolinaire M, Louis Z (1997) Carbon dynamics in slash-and-burn agriculture and land use alternatives of the humid forest zone in Cameroon. Agric Ecosyst Environ 65(3):245–256. https://doi.org/10.1016/s0167-8809(97)00060-1

Lal R, Follett RF, Kimble J, Cole CV (1999) Managing U.S. cropland to sequester carbon in soil. J Soil Water Conser 54(1):374–381

Li G (2004) Estimation of Chinese terrestrial net primary production using LUE model and MODIS data. Beijing: Institute of Remote Sensing Applications Chinese Academy of Sciences, pp 1–153

Lieth H (1972) Modelling the primary productivity of the world. Nat Resour 8(2):5–10

Lieth H, Box E (1972) Evapotranspiration and primary productivity: C. W. Thornthwaite memorial model. Publ Climatol 25(2):37–46

Lieth H, Whittaker RH (1975) Primary productivity of the biosphere, vol 14, Ecological studies and synthesis vol Accessed from http://nla.gov.au/nla.cat-vn1831028. Springer, New York

Liu M (2011) Preliminary study on the climatic extremes and their impacts on grain yield over the past 50 years in China. Beijing: University of Chinese Academy of Sciences, pp 31–43

Liu JY, Liu ML, Zhuang DF, Zhang ZX, Deng XZ (2003) Study on spatial pattern of land-use change in China during 1995-2000. Sci China Ser D Earth Sci 46(4):373–384

Liu JY, Zhang ZX, Xu XL, Kuang WH, Zhou WC, Zhang SW, Li RD, Yan CZ (2009) Spatial patterns and driving forces of land use change in China in the early 21st century. Acta Geograph Sin 64(12):1411–1420

Liu Y, Quan W, Gao Y (2010) Net primary production and its spatio-temporal pattern in North China. J Nat Resour 25(4):564–573

Parton WJ, Rasmussen PE (1994) Long-term effects of crop management in wheat-fallow: II, CENTURY model simulations. Soil Sci Soc Am J 58(2):530–536

Parton WJ, Schimel DS, Cole CV, Ojima DS (1987) Analysis of factors controlling soil organic matter levels in Great Plains grasslands. Soil Sci Soc Am J 51(5):1173–1179. citeulike-article-id:6695029

Pathak H, Byjesh K, Chakrabarti B, Aggarwal PK (2011) Potential and cost of carbon sequestration in Indian agriculture: estimates from long-term field experiments. Field Crops Res 120(1):102–111. https://doi.org/10.1016/j.fcr.2010.09.006

Piao S, Fang J, Guo Q (2001) Application of CASA model to the estimation of Chinese terrestrial net primary productivity. Acta Phytoecol Sin 25(5):603–608

Piao S, Fang J, Ciais P, Peylin P, Huang Y, Sitch S, Wang T (2009) The carbon balance of terrestrial ecosystems in China. Nature 458(7241):1009–1013

Piao S, Ciais P, Lomas M, Beer C, Liu H, Fang J, Friedlingstein P, Huang Y, Muraoka H, Son Y, Woodward I (2011) Contribution of climate change and rising CO2 to terrestrial carbon balance in East Asia: a multi-model analysis. Glob Planet Chang 75(3-4):133–142

Potter CS, Randerson JT, Field CB, Matson PA, Vitousek PM, Mooney HA, Klooster SA (1993) Terrestrial ecosystem production: a process model based on global satellite and surface data. Global Biogeochem Cycles 7(4):811–842. citeulike-article-id:2056763

Potter KN, Torbert HA, Jones OR, Matocha JE, Morrison JE, Unger PW (1998) Distribution and amount of soil organic C in long-term management systems in Texas. Soil and Tillage Research 47(3-4):309–321.

Riedel A, Michel C, Gosselin M, LeBlanc B (2008) Winter-spring dynamics in sea-ice carbon cycling in the coastal Arctic Ocean. Journal of Marine Systems 74 (3-4):918–932.

Schneider UA (2000) Agricultural options for mitigation of greenhouse gas emissions in the United States. Texas: Texas A&M University, pp 1–224

Schneider UA, McCarl BA, Schmid E (2007) Agricultural sector analysis on greenhouse gas mitigation in US agriculture and forestry. Agric Syst 94(2):128–140

Smith B, Knorr W, Widlowski J-L, Pinty B, Gobron N (2008) Combining remote sensing data with process modelling to monitor boreal conifer forest carbon balances. For Ecol Manag 255(12):3985–3994. https://doi.org/10.1016/j.foreco.2008.03.056

Song KS, Liu DW, Wang ZM, Zhang B, Jin C, Li F, Liu HJ (2008) Land use change in Sanjiang Plain and its driving forces analysis since 1954. Acta Geograph Sin 63(1):93–104

Sun R, Zhu Q (2000) Distribution and seasonal change of net primary productivity in China from April, 1992 to March, 1993. Acta Geograph Sin 55(1):36–45

Svirejeva-Hopkins A, Schellnhuber HJ, Pomaz VL (2004) Urbanised territories as a specific component of the Global Carbon Cycle. Ecol Model 173(2-3):295–312. https://doi.org/10.1016/j.ecolmodel.2003.09.022

Tian W, Xu K, Bing X, Zhai J, Zhang Z, Cheng Z, Wu Z (2007) Study on some agronomic traits of soybean cultivars with year of release in Jilin province. Chinese J Oil Crop Sci 29(4):397–401

Trumbore SE, Chadwick OA, Amundson R (1996) Rapid exchange between soil carbon and atmospheric carbon dioxide driven by temperature change. Science 272:393–396. https://doi.org/10.1126/science.272.5260.393

Uchijima Z, Seino H (1988) An agroclimatic method of estimating net primary productivity of natural vegetation. J Agric Res 21(4):244–250

Usuga JCL, Toro JAR, Alzate MVR, de Jesús Lema Tapias Á (2010) Estimation of biomass and carbon stocks in plants, soil and forest floor in different tropical forests. For Ecol Manag 260(10):1906–1913. https://doi.org/10.1016/j.foreco.2010.08.040

Wang Y, Huang Y, Zhang W, Yu Y, Wang P (2006) Simulating net primary production of agricultural vegetation in China (II): model validation and estimation of net primary production. J Nat Resour 21(6):916–925

West T, Marland G (2002) A synthesis of carbon sequestration, carbon emissions, and net carbon flux in agriculture: comparing tillage practices in the United States. Agric Ecosyst Environ 91(1):217–232

Xiao JY, Shen YJ, Ge JF, Tateishi R, Tang CY, Liang YQ, Huang ZY (2006) Evaluating urban expansion and land use change in Shijiazhuang, China, by using GIS and remote sensing. Landsc Urban Plan 75(1-2):69–80

Xue L (2004) Spatio – temporal of crop evapotranspiration and yield based on NDVI in North China Plain from 1981 to 2001. Beijing: China Agricultural University, pp 1–46

Yoshimura T, Nishioka J, Suzuki K, Hattori H, Kiyosawa H, Watanabe YW (2010) Impacts of elevated CO2 on organic carbon dynamics in nutrient depleted Okhotsk Sea surface waters. J Exp Mar Biol Ecol 395(1-2):191–198. https://doi.org/10.1016/j.jembe.2010.09.001

Yu Y, Guo Z, Wu H, Kahmann JA, Oldfield F (2009) Spatial changes in soil organic carbon density and storage of cultivated soils in China from 1980 to 2000. Glob Biogeochem Cycles 23(2):GB2021

Yuan W, Guan C (1997) Harvest index in rapessd affected by a few physiological factors. Acta Agron Sin 23(5):580–586

Yuan W, Guan C, Liao A (1999) Contribution of harvest index to seed yield of rapeseed. J Nat Sci Hunan Normal Univ 22(1):65–69

Zhang X (1999) Crop roots and soil water utilization. China Meteorological Press, Beijing, pp 35–51

Zhou G, Zheng Y, Chen S, Luo T (1998) NPP model of natural vegetation and its application in China. Sci Silvae Sinicae 34(5):2–11

Zhu HY, He SJ, Zhang M (2001) Driving forces analysis of land use change in Bohai Rim. Geogr Res 20(6):669–678

Changes in Nitrogen Application and Conservation Reserve Program Area from Cellulosic Biofuel Production in the United States

Jerome Dumortier

Abstract Biomass production from agricultural residues and dedicated bioenergy crops to meet the cellulosic biofuel mandate will change the land-allocation in the United States as well as the application rate of nitrogen. We extend a previously used model in three ways: First, we incorporate land in the Conservation Reserve Program (CRP) as a potential source for bioenergy crops. Second, we include high removal coefficients for agricultural residue removal and lastly, we assess changes in the nitrogen load at the county level. We quantify those effects under various scenarios differing by bioenergy crop (switchgrass and miscanthus), production costs, agricultural residue removal rates, and biofuel mandates. Consistent with previous research, our results indicate that little CRP land is converted to bioenergy crop production which is due to relatively high CRP payments compared to biomass yields. In addition, the majority of the cellulosic biofuel mandate can be met with agricultural residues which require more nitrogen in the corn and wheat producing regions in the U.S., i.e., the Midwest and the Great Plains, due to nutrient replacement. Our findings can inform future research to assess the effects of spatially shifting nitrogen application at the watershed level due to cellulosic biomass production.

Keywords Conservation reserve program · CRP · Greenhouse gas emissions · Biomass yield · Switchgrass · Miscanthus

1 Introduction

An important aspect of future agricultural production is the potential to produce energy from biomass. The U.S. Environmental Protection Agency (EPA) has currently a standard for advanced biofuels from cellulosic feedstock that requires

J. Dumortier (✉)
School of Public and Environmental Affairs, Indiana University – Purdue University Indianapolis, Indianapolis, IN, USA
e-mail: jdumorti@iupui.edu

© Springer International Publishing AG, part of Springer Nature 2018   189
R. Li, A. Monti (eds.), *Land Allocation for Biomass Crops*,
https://doi.org/10.1007/978-3-319-74536-7_10

the production of 60×10^9 liters (L) by 2022 (EISA, 2007). Over the past decade, the U.S. has derived ethanol mostly from corn and around 40% of current corn production is used for this purpose. In the future, a variety of scenarios can be envisioned. First, agricultural residues from corn, wheat, and sorghum can be used to produce cellulosic ethanol. Using agricultural residues has the advantage of limited ability to distort food and feed markets. Second, dedicated bioenergy crops such as switchgrass and miscanthus – that can also be grown on marginal land – can be used to produce cellulosic ethanol as well. Dedicated biomass crops have a higher yield potential but are also more expensive to produce. In addition, they compete directly for land with crops used for food and feed production. For both, agricultural residues and dedicated bioenergy crops, it is also possible to use them directly for bioelectricity production instead of liquid transportation fuels which is shown to be more effective to reduce greenhouse gas (GHG) emissions (Adler et al., 2007; Campbell et al., 2009). From a global perspective, Wang et al. (2012) argue that cellulosic feedstocks such as agricultural and forest residues, hold significant potential based on their availability. Given that biomass and bioenergy will most likely play an important role in the future to meet energy needs, it is important to evaluate the environmental effects in terms of greenhouse gas (GHG) emissions, ecosystem services provisions, and water quality. This analysis sheds light on the last two issues by incorporating effects on land in the Conservation Reserve Program (CRP) which provides ecosystem services and nitrogen application that affects water quality by extending a previously developed county-level economic model (Dumortier, 2013, 2016).

An increase in the production of biomass derived from either agricultural residues or dedicated bioenergy crops will lead to a change in land-allocation across the United States as well as a change in management practices (Dumortier, 2016). For example, collecting some of the agricultural residues may require an increase in nitrogen, phosphorus, and potassium use to compensate for the loss of nutrients associated with the removal of agricultural residues. Moreover, growing switchgrass or miscanthus instead of conventional crops such as corn or soybeans will lead to different nitrogen application rates than conventional cropping systems. Previous literature has assessed the nitrous oxide emissions from bioenergy crop production and the implications in terms of life-cycle analysis (Adler et al., 2007; Searchinger et al., 2008; Hudiburg et al., 2015). In this analysis, we will assess the spatial shift of (1) nitrogen application due to the collection of agricultural residues and dedicated bioenergy crop production and (2) the implications for the Conservation Reserve Program (CRP) in the contiguous United States. CRP land provides ecosystem services, improvements in water quality, and carbon sequestration among other benefits (Gelfand et al., 2011). The results of our analysis will have implications on future policies with respect to water quality and provision of ecosystem services. Although our model assess land-use and nitrogen load at the county level, it can help determine and evaluate the future nitrogen loads in watersheds from different bioenergy policies. For example, if nitrogen application rates change as a result of biomass production, then this will have long-term implications on nitrogen leaching and runoff for watersheds. In addition, we determine how much CRP land will be converted to accommodate different cellulosic biofuel mandates.

The land-use component of the model is extended by adding CRP land as an additional land-use category and by evaluating the change in nitrogen application. To answer our research questions, we first establish a baseline without any cellulosic biofuel mandate and then compare the baselines to various scenarios. The scenarios differ in terms of bioenergy crop (switchgrass or miscanthus), production cost (low and high), and agricultural residue removal rate (low and high). Besides assessing the spatial shift of nitrogen application and CRP land, we also assess the impact on commodity prices for corn, soybeans, and wheat.

2   Literature Review

There are multiple interrelated aspects involved with collecting agricultural residues or growing perennial grasses for bioenergy production. From an economic perspective, growing switchgrass or miscanthus will change the land-use allocation decision by the farmer and thus, has consequences on commodity supply and prices. From the environmental and agronomic standpoint, issues of water quality, biodiversity, ecosystem services provision, soil carbon changes, etc. arise. What sparked the interest in bioenergy production is its ability to contribute to energy security and GHG emission reduction compared to gasoline and diesel. Thus, GHG emissions associated with a change in management systems and cropping practices and type are an integral part of the policy debate.

Adler et al. (2007) conduct a life-cycle analysis for corn, soybean, alfalfa, hybrid poplar, reed canary grass, and switchgrass as bioenergy crops in Pennsylvania. The authors use the biogeochemical model DAYCENT to quantify GHG fluxes from the soil under various production scenarios. The authors find that the major contributor to GHG emissions from bioenergy feedstocks are nitrous oxide emissions partially caused by the application of nitrogen fertilizer. They find that hybrid poplar and switchgrass provide the largest reduction of GHG emissions compared to gasoline (approximately 115%) followed by reed canary grass (approximately 85%) and corn ethanol (40%). Similar results are found by Wang et al. (2012) but with miscanthus achieving an even higher reduction compared to switchgrass. Adler et al. (2007) do not include emissions from indirect land-use change that were included in subsequent research by Searchinger et al. (2008) and Fargione et al. (2008). They find that including emissions triggered by land-use change significantly decreases the GHG benefit from corn ethanol. Later studies have found mitigating factors with respect to indirect land-use change such as market mediating effects (Hertel et al., 2010) and yield growth (Dumortier et al., 2011) but the issue remains a major discussion point for corn ethanol. Kim et al. (2009) cite the possibility of no-till and cover crops as a possibility to improve the GHG balance of conventional corn ethanol. Wang et al. (2012) analyze life-cycle emissions from corn, sugarcane, corn stover, switchgrass, and miscanthus and find a decrease in energy use and GHG emissions when switching from corn to sugarcane and bioenergy crops. Davis et al. (2012) assess differences in GHG emissions, soil carbon (C) stock changes, and nitrogen leaching of replacing corn area with perennial grasses, i.e., switchgrass and miscanthus, in the U.S. Midwest. The removal of corn residues can decrease the soil organic carbon level and can increase GHG emissions associated with biofuels (Liska et al., 2014). The authors find that they could add 50–70 g CO_2 MJ^{-1} and would likely result in cellulosic biofuel not being able to meet the regulatory requirements set by U.S. legislation. The common thread among those studies is that cellulosic biofuels are better than conventional biofuel from a life-cycle GHG perspective but that there is uncertainty surrounding the exact benefit. This analysis forms the basis to get a better understanding about the effects nitrogen addition will have at the county-level. Given the volume of cellulosic ethanol, there may be a non-linear effect in terms of emissions because agricultural residues are used/harvested before farmers switch to perennial grasses.

Effects of bioenergy feedstock production on land-use and especially CRP land has been analyzed using economic models. The current availability of agricultural residues does not meet the entire cellulosic biofuel mandate 60×10^9 L (only about one-third in the case of corn stover) and the deficit needs to be covered by other sources such as perennial grasses or woody biomass (Khanna et al., 2010; Perlack and Stokes, 2011). Khanna et al. (2010) analyze the production of biomass in Illinois and show that under the mandate, total land allocated to agricultural production increases by 5%. Their results indicate that land share in corn increases 6–8 percentage points and the land share for soybeans, wheat, and pasture decreases by 16, 15, and 44 percentage points, respectively. At least in the case of Illinois, switchgrass is not produced due to its cost disadvantage compared to miscanthus. Increases in nitrogen is caused by (1) the expansion of corn acres, (2) the change from a corn-soybean to a continuous corn rotation, and (3) nitrogen additions to replace nutrients after corn stover removal. Using the POLYSYS framework, Hellwinckel et al. (2016) find that only a minimal amount of 0.29 million hectares (3%) of CRP land is converted to biomass production. Crop residues cover 78% of the mandate and woody biomass covers 19% of the mandate. The authors do not find a conversion of CRP land to dedicated biomass crops.

Gelfand et al. (2011) estimate the carbon balance of CRP land converted to cropland or perennial grasses by including changes in agronomic inputs and carbon stock, forgone carbon sequestration as well as fluxes of methane and nitrous oxide. Under the no-tillage scenario, the emissions are up 68 t CO_2-e ha^{-1} whereas under the tillage scenario, carbon emissions are up to 222 t CO_2-e ha^{-1}. Qin et al. (2015) estimates emissions from the production of switchgrass and miscanthus if grown on current corn-producing land in the United States. Emissions from nitrogen application are important for corn production and contribute over 90% of GHG emissions from corn whereas switchgrass and miscanthus systems are found to be a carbon sink. Implemented conservation practices to improve water quality or to reduce negative impacts from nitrogen leaching and run-off such as the Gulf of Mexico hypoxic zone will be impacted by the location of nitrogen application (Rabotyagov et al., 2014). This becomes important since Rabotyagov et al. (2014) estimate that a reduction of the hypoxic zone to 5,000 km^2 is achieved at a cost of 2.7×10^9 annually. So efforts with respect to conservation policies and cellulosic biomass production need to be analyzed simultaneously to avoid inefficient outcomes or adverse consequences. This is a similar conclusion reached by Gramig et al. (2013) who assess the trade-offs associated with corn stover collection under various rotation and tillage scenarios in terms of cost, stover yield, nitrates, phosphorus, sediments, and global warming potential. Stover collection increases the GHG emissions but reduces nitrate and phosphorus loads. The authors suggest that more information is needed with respect to nutrient replacement in order to maintain crop yields.

3   Economic Model

Our economic model is closely based on Dumortier (2013, 2016) and we focus on the area of corn, soybeans, wheat, pasture and CRP land for biomass feedstock production. The model covers the contiguous United States and includes corn stover, wheat straw, switchgrass and miscanthus as possible bioenergy feedstock. Counties producing at least one of the field crops are included in the analysis. The model is based on Dumortier (2016) but is extended in three ways: (1) CRP land is available for biomass production, (2) high residue removal rates are incorporated, and (3) nitrogen emissions are calculated as an output of the model.

We assume a profit-maximizing, representative landowner in each county who determines the allocation of land to field crops, pasture, CRP, and dedicated bioenergy crops. In addition, the landowner can harvest agricultural residues from corn and wheat. The biomass price is measured in $ GJ^{-1} and the prices for corn, soybean, and wheat are endogenous to the model. During the simulation exercise, we impose a cellulosic biofuel mandate and the model solves for the biomass and commodity prices that fulfill the binding biofuel mandate. We denote the field crops with the subscript $j \in \{1, 2, 3\}$ and the landowners/counties with subscript $i \in \{1, \ldots, N\}$ where N represents the total number of counties. The superscripts f, r, s, and c denote variables and parameters associated with field crops, agricultural residues, bioenergy crops, and CRP land, respectively. The national prices for crops and biomass are p_j and p_b, respectively.

For the three field crops, the landowner faces demand function $Q_j = D(p, e)$ for crop j which is a function of the price vector $p \in \{1, 2, 3\}$ and the conventional biofuel demand e. For each crop, there are three demand sectors, i.e., consumer/food, feed, and export. The demand for each sector $m \in \{1, 2, 3\}$ in our model is written as:

$$Q_j = D(p, e) = \sum_{m=1}^{M} \left[\gamma_{jm} \prod_{j=1}^{J} p_j^{\theta_{jm}} \right] + e \tag{1}$$

where γ_{jm} are the constants and θ_{jm} are interpreted as own-price and cross-price elasticities. We only include corn ethanol as the conventional biofuel in our model.

Landowner i has seven decision variables for the land allocation: three field crops, land harvested for agricultural residues, CRP, and dedicated bioenergy crops. Similar to Dumortier (2016), the profit from field crops $B_i^f(\cdot)$ not accounting for the returns from agricultural residue collection is expressed as:

$$B_i^f(a_{ij}^f, a_{ij}^r) = \sum_{j=1}^{3} p_j \left(a_{ij}^f + a_{ij}^r\right) y_{ij} - \sum_{j=1}^{3} C_{ij}(a_{ij}^f, a_{ij}^r) \tag{2}$$

where the areas allocated to corn, soybeans, and wheat not harvested for agricultural residues are denoted by a_{ij}^f. As aforementioned, the advantage of harvesting agricultural residues for the production of cellulosic biofuel is that it does not interfere with supplying field crops. Thus, the cost function for field crops includes the area harvested for residues a_{ij}^r as well. The cost function $C_{ij}(\cdot)$ can be written as:

$$C_{ij}(a_{ij}^f, a_{ij}^r) = \alpha_{ij} \left(a_{ij}^f + a_{ij}^r\right) + \frac{1}{2}\beta_{ij} \left(a_{ij}^f + a_{ij}^r\right)^2 \tag{3}$$

where $\alpha_{ij} > 0$ and $\beta_{ij} > 0$ are county and crop specific cost parameters. Note that $\partial C_{ij}(\cdot)/\partial a_{ij}^f > 0$ and $\partial C_{ij}(\cdot)/\partial a_{ij}^r > 0$ and thus, the marginal cost is increasing capturing either decreasing yields on marginal land or the necessity to use more nitrogen for the same reason (Mallory et al., 2011). The profits from residues are written as:

$$B_i^r(a_{ij}^r) = p_b \sum_{j=1}^{3} \delta_{ij} y_{ij} a_{ij}^r - \sum_{j=1}^{3} \eta_{ij} a_{ij}^r \tag{4}$$

where δ_{ij} summarizes the agricultural residue specific energy content and the county-specific residue removal coefficient. Equation (4) assumes that harvesting agricultural residues results in a fixed per hectare cost and does not influence marginal cost. There are two issues that need to be highlighted with respect to this assumption: First, we are guaranteed to have a solution to the land allocation problem of the farmer because equation (3) is non-linear. Second, the specifications result in an all-or-nothing allocation of land to residue collection in county i. The profit from bioenergy crop production is written as:

$$B_i^s(a_i^s) = p_b \delta_i^s y_i^s a_i^s - \eta_i^s a_i^s \tag{5}$$

Note that for land in the CRP, we assume that there are no operating costs and thus, the landowner just receives a rental payment that we denote $B_i^c(a_i^s)$. Given the profit from field crops, residue collection, bioenergy crops, and CRP land, the profit maximization for the landowner can be represented by the sum of equations (2), (4), (5), and the rental payment for CRP land i.e.,

$$B(a) = B_i^f(a_{ij}^f, a_{ij}^r) + B_i^r(a_{ij}^r) + B_i^s(a_i^s) + B_i^c(a_i^c) \tag{6}$$

subject to a binding land constraint and non-negativity constraints. Setting up the Lagrangian and deriving the following first order conditions for a_{ij}^f, a_{ij}^r, a_i^s, and a_i^c is straightforward.

The simulation model imposes an exogenous cellulosic biofuel mandate that needs to be met and we numerically solve for the equilibrium that meets the mandate and clears the commodity markets based on the demand equation (1) and the aggregate production of farmers. The commodity and biomass prices are determined endogenously in our model based on the first order conditions, the demand equations, and the cellulosic biofuel mandate. Based on the biomass price, we find a competitive equilibrium such that all markets clear. The numerical procedure will start with the given biomass price plus the three crop prices representing starting values. Based on the prices, landowners allocate their land and crop as well as biomass production emerges, if the resulting production is not consistent with the demand and the mandate at those prices, the algorithm continues until convergence. Agriculture is a perfectly competitive market and hence, all agents are price takers and do not take the effect of their acreage decision on output prices into account. Due to differences in the energy content of corn stover, wheat straw, and bioenergy crops, we express the biomass price in dollars per gigajoule (GJ).

4   Data

The five data components necessary to conduct out analysis are: (1) commodity demand, (2) field crop area, yield, and cost (3) biomass yields and cost for agricultural residues and switchgrass/miscanthus, (4) nitrogen application, and (5) CRP land and payments. In addition, we outline the scenarios that are analyzed in the results section.

4.1   Commodity Demand

The baseline model in the absence of any biomass production is calibrated to 2022 in terms of commodity prices, demand, yield, and area. The prices and price elasticities for food, feed, and export for demand equation (1) can be found in Table 1 and are taken from FAPRI (2011a) and Chen (2010). We use the 2022 crop prices and demand quantities from FAPRI (2016) to determine the constant (γ_{jm}) as well as the use of 5.418×10^9 bushels (or 137.66 million metric tons) of corn for ethanol production. The commodity prices represent averages from FAPRI (2016) over the period from 2015 to 2022. We hypothesize that the level of commodity prices observed does not affect our prices significantly. The price level affects when farmers decide to either harvest agricultural residues besides the regular harvest or change to bioenergy crops but does not change the "sequencing" of counties

Table 1 Prices and price elasticities for food, feed, and export. All elasticities are from FAPRI (2011b) with the exception of food/consumer demand for corn and export demand for soybeans which are taken from Chen (2010). The base prices are deflated to 2012 Dollars using the producer price index

	γ_{jm}	p_{CO}	p_{SB}	p_{WH}
Base price ($ bu^{-1})		3.60	8.98	4.93
Base price ($ t^{-1})		141.73	329.96	181.15
Food/consumer demand				
Corn	114.03	−0.230	−	−
Soybeans	626.77	−	−0.434	−
Wheat	54.70	−	−	−0.075
Feed demand				
Corn	53.01	−0.201	−	−
Exports				
Corn	550.20	−0.570	−	0.120
Soybeans	1347.44	0.030	−0.63	0.020
Wheat	5727.43	0.170	0.040	−1.230

that change to biomass collection since that decision is mainly driven by the crop yields of the county relative to the yields of biomass collection. The calibration of the demand is similar to Dumortier (2013, 2016) with the exception that the price and demand data was updated to take into account the recent decrease in commodity prices.

4.2   Field Crop Production

The county-level area harvested is obtained from the National Agricultural Statistics Service (NASS) for 2008–2012. Yield and area for crop j in county i was set to zero if crop production occurred in less than 2 years between 2000 and 2010. The base area a_{ij} is the average crop area harvested between 2000 and 2010 corrected to match total domestic use and exports in 2022. With no price on biomass, the areas correspond to a_{ij}^{f} in Eq. (2). We use the 2022 expected yields by crop and county from the University of Missouri Food and Agricultural Policy Research Institute Farm Cost and Return Tool (FAPRI CART) to determine the yield parameter y_{ij} used in our simulation.

Besides the expected yield, FAPRI CART provides cost and return classified according to the USDA Economic Research Service Farm Resources Regions. We proceed in two steps to obtain county specific parameters α_{ij} and β_{ij}. First, we use the FAPRI CART parameters on operating cost by crop and farm resource region for 2022 and set the parameter α_{ij} equal to the total operating cost composed of chemicals, custom operations, fertilizer, energy (i.e., fuel, lube, and electricity), interest on operating capital, repairs, seeds, and miscellaneous costs. If acreage is increased, we assume that more inputs are needed per hectare resulting in increasing marginal cost. The increase in marginal cost might be due to a necessary increase in fertilizer use or a reduction in yields because of decreasing land quality (Mallory et al., 2011). We assume that all counties in a particular farm resource region have the same α_{ij}. The values are represented in the supplemental material. Second, assuming profit maximizing but price taking behavior allows the calculation of the county specific parameters β_{ij} because the landowner sets marginal revenue equal to marginal cost, i.e., $p_j \cdot y_{ij} = \alpha_{ij} + \beta_{ij} a_{ij}^f$. Given p_j, y_{ij}, α_{ij}, and a_{ij}^f enables us to obtain β_{ij} for the base year. The projection of the yield in FAPRI CART is based on a linear projection. We do not account for possible climate change impacts since those are probably negligible over the projection period considered. Climate change impacts should probably be considered for a projection model in the long-run.

4.3 Biomass Production

The most important component in our model is the biomass yield potential at the county level as well as the cost associated with biomass harvesting. We differentiate scenarios in our model along three dimension related to biomass: (1) bioenergy crop, i.e., switchgrass versus miscanthus, (2) high versus low production cost, and (3) low versus high residue removal coefficients. Especially the inclusion of high residue removal coefficient is an important extension of the model presented in Dumortier (2016).

The switchgrass and miscanthus yields were obtained from Miguez et al. (2012) who calculated yield potential based on field observations in the United States. Khanna et al. (2010) assume no yield loss during the harvest operations of switchgrass but a 33% loss of peak miscanthus yield during harvest. We assume a lifespan of 10 and 15 years for switchgrass and miscanthus, respectively. As in previous literature, we annualize all the costs to obtain yearly operating cost (Perrin et al., 2008; Jain et al., 2010). We calculate the net present value of all the operating cost by assuming a discount rate of 4% and then annualize the net present value as follows:

$$\text{Annual Cost} = r \cdot NPV \cdot \left(1 - \frac{1}{(1+r)^t}\right)^{-1}$$

With respect to the biomass production cost, we use data from Chen (2010) who provides state-level cost on mowing, raking, and bailing operations. Upper and lower bounds are reported for those harvest costs. For the case of agricultural residues, only raking and bailing costs are considered whereas the harvest of bioenergy crops also includes mowing.

Considerable discussion occurs about the sustainable percentage of agricultural residues that can be removed after the regular harvest operation. Perlack and Stokes (2011) presents two levels of removal coefficients at the county level, i.e., high and low. In this analysis, we assume that both removal levels do not interfere with commodity yields and can be compensated by a higher application of fertilizer. We assume that the nutrient[1] replacement rates (in kilogram per metric ton of residue removed) are 7.95 (N), 2.95 (P), and 15 (K) for corn stover and 6.00, 1.58, and 10.75 for wheat straw.

4.4 Nitrogen Application Rates

Besides the inclusion of high removal coefficients for agricultural residues, this analysis also includes the changes in the nitrogen application at the county level as an extension to previous models. There is little to no data about the application rates per county and thus, some assumptions need to be made. The USDA Economic Research Service provides data about fertilizer use and prices for selected crops, states, and years (USDA Economic Research Service, 2011). Throughout our analysis, we assume that the fertilizer application is driven by the yield, i.e., the higher the yield the more nitrogen is necessary (Khanna et al., 2010).

In a first step, we take the fertilizer application rate of the most recent year available by state and crop and divide it by the corresponding expected state yield to obtain an application rate per ton of yield (Table 2). We use the expected yield rate instead of the actual yield rate to assure that yield the application rate is not inflated or deflated by an unusual yield occurrence. For states with missing application rates, we use the values of neighboring or close by states. To assess the sensitivity of our results with respect to nitrogen application, we also use a simple ordinary least squared (OLS) model to regress the N application rate on expected yield based on data provided the USDA's Economic Research Service (USDA Economic Research Service, 2011). The data includes the application rate for selected crops, states, and years. We include states that result in a positive slope ranging from 6.38 to 14.40 kg N per ton of yield from a base application of fertilizer. The R^2 for the states included is above 0.56 with the exception of Missouri (0.23) and Pennsylvania (0.33). The resulting differences between the values presented in Table 2 and the OLS model are summarized by county in Fig. 1. Note that farmers tend to over apply nitrogen in order to hedge against excessive rain (Rosas et al., 2015). The estimated U.S. plant nutrient use in 2010 for the three crops was 6.4 million metric tons (USDA Economic Research Service, 2011). Our baseline estimates for 2022 estimates are 7.95 and 7.86 million metric tons assuming that the difference is due to increases in yield.

With respect to switchgrass, we use an application rate of 56 and 140 kg ha⁻1 in the low and high cost scenario, respectively. For miscanthus, those values are 50 and 130 kg ha⁻1. Qin et al. (2015) assumes four different nitrogen application

---

[1]Nitrogen (N), Phosphate (P), and potassium (K)

Table 2 Minimum, median, and maximum N application rate per hectare by state for corn, soybeans, and wheat as well as the N application rate in kg per ton of yield

State	Corn			Soybean			Wheat			kg N per ton		
	min	med	max	min	med	max	min	med	max	CO	SB	WH
Alabama	53	142	234	24	36	57	63	122	203	21.65	19.13	37.60
Arkansas	52	174	231	12	35	51	44	91	129	16.90	14.18	24.84
Delaware	61	67	70	27	30	31	132	140	144	7.07	10.92	24.21
Georgia	44	157	317	22	36	70	66	133	228	21.65	19.13	37.60
Illinois	123	171	206	18	28	34	65	99	137	14.55	7.74	20.25
Indiana	122	166	196	16	21	26	59	112	156	14.82	6.00	21.97
Iowa	140	184	203	20	24	27	35	60	114	14.15	6.19	21.97
Kansas	81	132	257	9	17	32	56	75	121	16.54	6.49	27.03
Kentucky	67	149	193	15	27	33	66	130	190	16.39	8.84	28.36
Louisiana	39	112	233	6	10	18	64	125	181	17.86	4.66	37.88
Maryland	49	62	76	33	46	55	69	121	148	7.07	17.39	24.21
Michigan	58	142	178	7	14	17	44	89	135	14.97	4.76	20.10
Minnesota	55	197	225	6	14	18	56	90	136	16.37	4.67	29.51
Mississippi	65	177	303	7	12	21	103	121	264	21.65	5.43	51.48
Missouri	61	155	206	8	22	27	45	94	129	16.90	7.41	24.84
Nebraska	100	169	211	9	18	23	42	76	135	14.76	4.60	23.24
New Jersey	36	53	64	16	20	28	63	100	113	7.07	8.56	24.21
New York	26	64	79	22	34	57	49	99	136	7.07	11.45	24.21
North Carolina	74	165	300	14	25	74	15	92	122	25.12	11.45	24.21
North Dakota	46	141	208	2	12	22	46	81	133	18.43	5.12	33.48
Ohio	82	168	192	13	19	22	50	97	124	15.73	5.67	19.15
Oklahoma	60	94	279	15	39	94	56	73	111	16.54	23.01	33.97
Pennsylvania	69	90	129	24	37	45	35	68	96	11.48	12.06	16.35
South Carolina	61	137	212	16	26	43	40	78	136	25.99	14.41	24.21
South Dakota	30	127	190	3	11	15	48	119	196	14.84	3.98	32.36
Tennessee	124	249	330	15	25	33	57	111	188	27.82	9.42	28.36
Texas	28	98	323	19	41	111	22	55	122	17.86	20.66	28.89
Virginia	25	107	183	17	28	50	15	111	153	17.14	12.08	24.21
West Virginia	62	90	181	28	34	37	53	69	172	16.33	12.08	24.21
Wisconsin	59	119	155	9	19	25	53	154	200	11.90	6.06	32.36

rates of 0, 67, 134, and 246 kg N ha^{-1} yr^{-1}. The authors argue that at the highest level, crop growth is not hindered by the lack of nitrogen. The authors also report that the nitrogen application varies between 70–180 kg N ha^{-1} yr^{-1} with regional variability. Switchgrass yield responds to nitrogen application and reaches full potential at 67 kg N ha^{-1} yr^{-1} in the model used by (Qin et al., 2015).

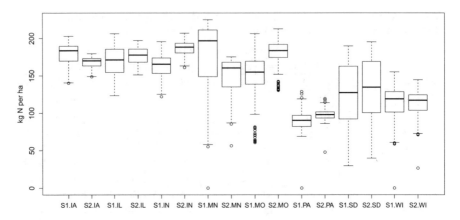

Fig. 1 Alternative corn N application rates (S2) compared to the baseline (S1) values for Iowa (IA), Illinois (IL), Indiana (IN), Minnesota (MN), Missouri (MO), Pennsylvania (PA), South Dakota (SD), and Wisconsin (WI)

4.5   Conservation Reserve Program Area and Payment

Data on annual enrollment and average payments by county for land in the CRP is provided by USDA Farm Service Agency.[2] We use 2014 enrollment and average county payments in our model. Chen and Li (2016) estimate the cost to maintain the 2007 level of CRP land in Illinois at different biomass prices. Depending on the scenario, maintenance costs range between 104.6–176.5 and 118.1–245.4 million for biomass prices of $50 and $150 respectively. For Illinois, they find conversion of a maximum of 0.445 million hectares at a biomass price of $150 dry ton^{-1} and low production cost. Throughout our analysis, we assume that the CRP payments do not change and we also do not impose any constraints with regard to re-enrollment, contract length, or minimum enrollment. The current payments in $ per hectare by state are depicted in Fig. 2.

4.6   Scenarios

To assess the effects of biomass production on nutrient application and CRP land allocation, we run scenarios differentiated by biomass production cost (high versus low), agricultural residue removal rates (high versus low), bioenergy crop (switchgrass versus miscanthus), and biofuel mandates ranging from 0 to 60×10^9 L. We evaluate a total of 33 scenarios including the baseline (Table 3). Since the

---

[2]https://www.fsa.usda.gov/programs-and-services/conservation-programs/reports-and-statistics/conservation-reserve-program-statistics/index

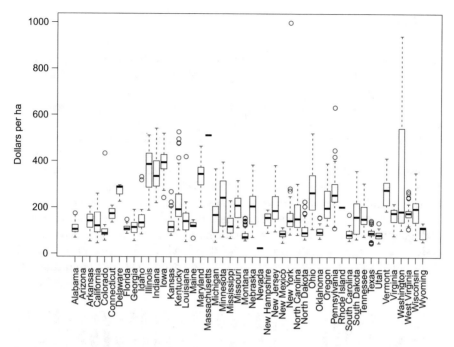

Fig. 2 Rental payments for CRP land in $ per ha by state

Table 3 Summary of scenarios differentiated by bioenergy crop, production cost, and agricultural residue removal rate. Besides the aforementioned differentiation, four cellulosic biofuel mandates are analyzed as well, i.e., 15, 30, 45, and 60 × 10^9 L

Scenario	Bioenergy crop	Production cost	Removal rate
Baseline	No biomass price		
SLRT 15–60	Switchgrass	Low	Reduced tillage
SHRT 15–60	Switchgrass	High	Reduced tillage
SLNT 15–60	Switchgrass	Low	No tillage
SHNT 15–60	Switchgrass	High	No tillage
MLRT 15–60	Miscanthus	Low	Reduced tillage
MHRT 15–60	Miscanthus	High	Reduced tillage
MLNT 15–60	Miscanthus	Low	No tillage
MHNT 15–60	Miscanthus	High	No tillage

2022 mandate calls for a production of 60 × 10^9 L of cellulosic ethanol, we focus on those scenarios when presenting our results. For each scenario, we assume that there is only one biomass crop available which allows to disentangle the effects of the two biomass crops on commodity prices, land allocation, and nitrogen use.

5   Results

Our results indicate that the share of perennial crops in cellulosic biofuel production is generally below 40% with the exception of one scenario, i.e., production of miscanthus under low cost and reduced tillage. Under high production cost and no-tillage, no bioenergy crops are produced even in the case of a required production of 60×10^9 liters. Dumortier et al. (2016) has shown that sufficient agricultural residues are available to cover the mandate under no-tillage removal coefficients (Fig. 3).[3] In the case of reduced tillage that requires a lower amount of residues removed, the production of cellulosic ethanol from bioenergy crops is necessary to meet the mandate. Since the cellulosic biofuel mandate requires the collection of agricultural residues, farmers have an incentive to increase the area of corn and wheat at the expense of soybeans. This results in the increase in the price of soybeans and a decrease in the prices of corn and wheat. Only in cases where agricultural residues are largely insufficient to cover the mandate do we see an increase in commodity prices such as in the miscanthus scenarios with low production cost and reduced tillage (Fig. 4). In all of our scenarios, the maximum change in CRP

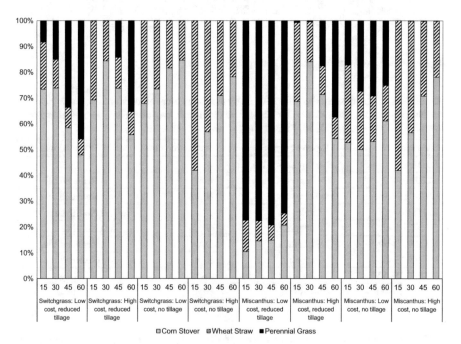

Fig. 3 Share of corn stover, wheat straw, and perennial grasses for the production of cellulosic biofuels at mandate levels of 15, 30, 45, and 60×10^9 L

---

[3]Although our model covers the contiguous U.S. for crop production, the biomass model (agricultural residues and bioenergy crops) are restricted to the Eastern parts of the country that are depicted in the figures.

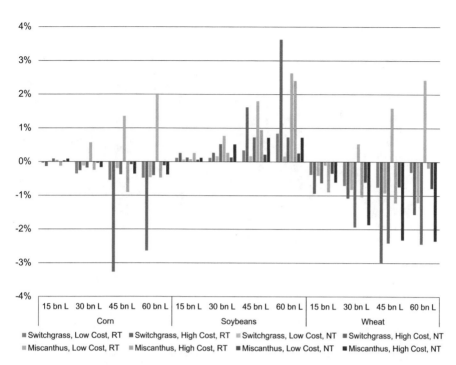

Fig. 4 Commodity price change from the baseline for corn, soybeans, and wheat under no-tillage (NT) and reduced tillage (RT)

land is below 0.6% which is slighty less than Hellwinckel et al. (2016) calculated. The median biomass price in our model is \$3.43 GJ^{-1} and the maximum biomass price is \$5.61 GJ^{-1}. Figure 5 illustrates that the CRP land payments are sufficiently high to preclude any conversion to bioenergy crops. Since the cellulosic biofuel mandate can be meet with the harvest of agricultural residues under the no-tillage scenarios, little to no bioenergy crops are grown. Our analysis shows that only under low biomass production cost and miscanthus do we see some minor conversion of cropland to miscanthus (Fig. 6). If farmers decide to switch to perennial grasses, the majority of them will be grown in the Southeast of the U.S. and in the Southern Great Plains. The conversion from cropland to bioenergy crops in the Great Plains is at the expense of wheat that yields low returns compared to corn and soybean. Crop area converted to switchgrass or miscanthus results in the loss of low yielding corn and soybean areas in the Southeast. The overall changes in nitrogen are represented in Fig. 7. The scenarios that see the highest overall increase in nitrogen use are in the case of no-tillage. The higher removal coefficients require more nitrogen to compensate for the nutrient loss in the soil. In scenarios that rely on agricultural residues such as no-tillage switchgrass and the no-tillage, high cost miscanthus scenarios, the increase in nitrogen use is linear to additional nitrogen requirements for nutrient loss. In scenarios that include either switchgrass or miscanthus, we see

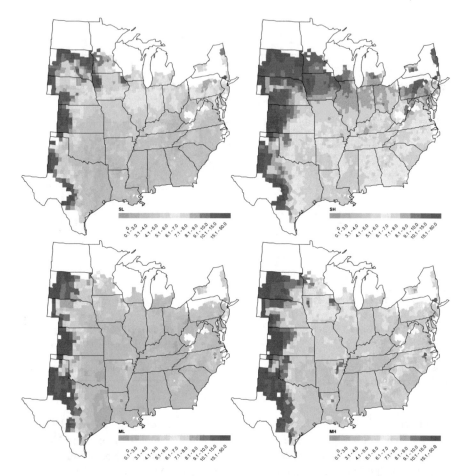

Fig. 5 Break-even biomass price necessary to trigger conversion of Conservation Reserve Program (CRP) land to dedicated bioenergy crops. The median break-even prices (per GJ) are for low cost switchgrass (SL), high cost switchgrass (SH), low cost miscanthus (ML), and high cost miscanthus (MH) are $4.53, $6.38, $3.64, and $5.08, respectively

initially an increase (due to an increase in the mandate) in the nitrogen consumption but at a decreasing rate since farmers are changing to perennial crops that require less nitrogen per acre than corn or wheat. The smallest increase in total nitrogen consumption can be observed in the scenario "Miscanthus: Low cost, reduced tillage" because it has the highest share of miscanthus (Fig. 8).

As aforementioned, the literature provides a range from one-third to 78% of the mandate being covered by agricultural residues (Khanna et al., 2010; Hellwinckel et al., 2016). As Khanna et al. (2010) point out, the important factor that determines the share of crop residues in covering the mandate are the assumptions made with respect to the residue removal rates. This is reflected in Fig. 3 which illustrates a wide range of agricultural residue shares to cover the mandate.

Fig. 6 Share of switchgrass and miscanthus under the 60×10^9 L scenarios. In order to facilitate the comparison with the figure that includes the changes in nitrogen application, we have also included the scenarios where no perennial crop is grown

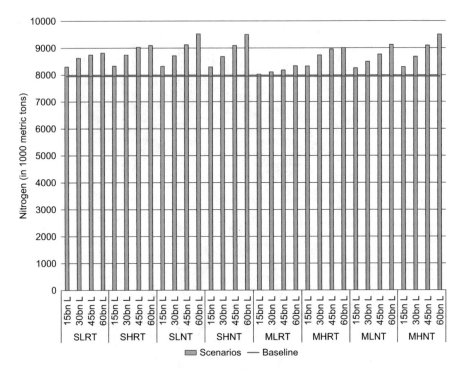

Fig. 7 Change in total nitrogen consumption with scenarios corresponding to Table 3

6 Discussion and Conclusion

Given a binding cellulosic biofuel mandate and various scenarios how to achieve the mandate, it is of interest to stakeholders what the effects on land allocation, nitrogen use, and CRP land are. In this paper, we have extended a previously used model in three ways: First, we included CRP land as an additional land area for bioenergy crops. Second, we incorporated the possibility of high residue removal rates that are still sustainable but require more nitrogen inputs to compensate for the nutrient loss. And lastly, we incorporated baseline nitrogen application rates by county for corn, soybeans, and wheat and assessed the changes that will occur with the harvesting of agricultural residues or bioenergy crops. We use an economic model for the contiguous U.S. that allows to assess possible non-linear effects on outcome variables from an increase in biomass production. For example, Dumortier (2016) shows that little to no dedicated bioenergy crops are grown for low levels of mandate because it can be covered from agricultural residues. As the mandate increase, dedicated bioenergy crops increase in production possibly leading to indirect land-use change. Our model is able to assess those effects because prices are modeled at the national level.

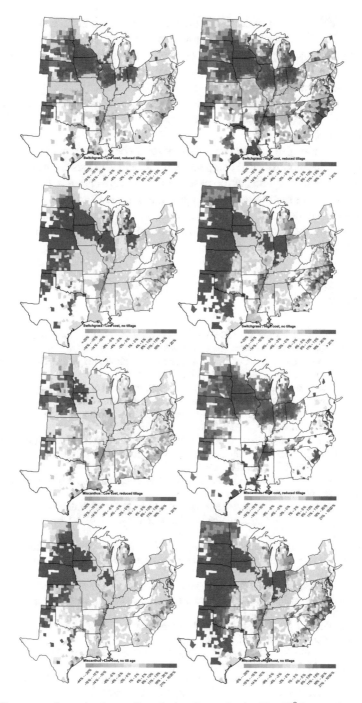

Fig. 8 Percentage change in nitrogen from the baseline under the 60×10^9 L scenarios

Very little CRP land is converted to bioenergy crops even under a biofuel mandate of 60×10^9 L. CRP payments are relatively high compared to the break-even price that makes the planting of bioenergy crops profitable. In general, little bioenergy crops are planted except under the assumption of reduced tillage/low removal coefficients. In that case, agricultural residues are not sufficient to cover the mandate and bioenergy crops are necessary to supplement as a feedstock for cellulosic fuels. Under high biomass production cost and high removal coefficients, no bioenergy crops are planted at all.

In terms of nitrogen application, high removal coefficients require a significant addition in terms of nitrogen to compensate for the nutrient loss. In the case of wheat straw, this leads to a significant increase in nitrogen application in the Great Plains. For corn stover, we see more nitrogen application in the Corn Belt. Future research needs to evaluate how this changes the nutrient loads in watersheds. Furthermore, our model does not include the price effects on an increase in nitrogen demand and how it affects the application rate.

References

Adler PR, Grosso SJD, Parton WJ (2007) Life-cycle assessment of net greenhouse-gas flux for bioenergy cropping systems. Ecol Appl 17(3):675–691

Campbell JE, Lobell DB, Field C (2009) Greater transportation energy and GHG offsets from bioelectricity than ethanol. Science 324(5930):1055–1057

Chen X (2010) A dynamic analysis of U.S. biofuels policy impact on land-use, greenhouse gases and social welfare. Ph.D. thesis, University of Illinois at Urbana-Champaign

Chen X, Li L (2016) Supply of cellulosic biomass in Illinois and implications for the conservation reserve program. Glob Change Biol Bioenergy 8(1):25–34

Davis SC, Parton WJ, Grosso SJD, Keough C, Marx E, Adler PR, DeLucia EH (2012) Impact of second-generation biofuel agriculture on greenhouse-gas emissions in the corn-growing regions of the US. Front Ecol Environ 10(2):69–74

Dumortier J (2013) Co-firing in coal power plants and its impact on biomass feedstock availability. Energy Policy 60:396–405

Dumortier J (2016) Impact of agronomic uncertainty in biomass production and endogenous commodity prices on cellulosic biofuel feedstock composition. Glob Change Biol Bioenergy 8(3):35–50

Dumortier J, Hayes DJ, Carriquiry M, Dong F, Du X, Elobeid A, Fabiosa JF, Tokgoz S (2011) Sensitivity of carbon emission estimates from indirect land-use change. Appl Econ Perspect Policy 33(3):428–448

Dumortier J, Kauffman N, Hayes DJ (2016) Production and spatial distribution of switchgrass and miscanthus in the united states under uncertainty and sunk cost. Working paper 16-WP 568, Iowa State University

EISA (2007) Public Law 110–140 Energy Independence and Security Act of 2007, Sec. 202. Renewable Fuel Standard (a)(1). Public law

FAPRI (2011a) FAPRI-MU Stochastic U.S. Crop Model Documentation. FAPRI-MU Report 09-11, Food and Agricultural Policy Research Institute

FAPRI (2011b) Model documentation for biomass, cellulosic biofuels, renewable and conventional electricity, natural gas and coal markets. FAPRI-MU Report 12-11, Food and Agricultural Policy Research Institute

FAPRI (2016) U.S. baseline briefing book: projections for agricultural and biofuel markets. FAPRI-MU Report 02-16, Food and Agricultural Policy Research Institute

Fargione J, Hill J, Tilman D, Polasky S, Hawthorne P (2008) Land clearing and the biofuel carbon debt. Science 319(5867):1235–1238

Gelfand I, Zenone T, Jasrotia P, Chen J, Hamilton SK, Robertson GP (2011) Carbon debt of Conservation Reserve Program (CRP) grasslands converted to bioenergy production. Proc Natl Acad Sci 108(33):13864–13869

Gramig BM, Reeling CJ, Cibin R, Chaubey I (2013) Environmental and economic trade-offs in a watershed when using corn stover for bioenergy. Environ Sci Technol 47:1784–1791

Hellwinckel C, Clark C, Langholtz M, Eaton L (2016) Simulated impact of the renewable fuels standard on US Conservation Reserve Program enrollment and conversion. Glob Change Biol Bioenergy 8:245–256

Hertel TW, Golub AA, Jones AD, O'Hare M, Plevin RJ, Kammen DM (2010) Effects of us maize ethanol on global land use and greenhouse gas emissions: estimating market-mediated responses. Bioscience 60(3):223–231

Hudiburg TW, Davis SC, Parton W, Delucia EH (2015) Bioenergy crop greenhouse gas mitigation potential under a range of management practices. Glob Change Biol Bioenergy 7:366–374

Jain AK, Khanna M, Erickson M, Huang H (2010) An integrated biogeochemical and economic analysis of bioenergy crops in the Midwestern United States. Glob Change Biol Bioenergy 2(5):217–234

Khanna M, Önal H, Chen X, Huang H (2010) Handbook of bioenergy economics and policy, vol 33. Natural resource management and policy. Springer, New York, pp 287–305. Chapter: Meeting biofuels targets: implications for land use, greenhouse gas emissions, and nitrogen use in Illinois

Kim H, Kim S, Dale BE (2009) Biofuels, land use change, and greenhouse gas emissions: Some unexplored variables. Environ Sci Technol 43:961–967

Liska AJ, Yang H, Milner M, Goddard S, Blanco-Canqui H, Pelton MP, Fang XX, Zhu H, Suyker AE (2014) Biofuels from crop residue can reduce soil carbon and increase CO_2 emissions. Nat Clim Change 4:398–401

Mallory ML, Hayes DJ, Babcock BA (2011) Crop-based biofuel production with acreage competition and uncertainty. Land Econ 87(4):610–627

Miguez FE, Maughan M, Bollero GA (2012) Modeling spatial and dynamic variation in growth, yield, and yield stability of the bioenergy crops Miscanthus x giganteus and Panicum virgatum accross the conterminous United States. Glob Change Biol Bioenergy 4:509–520

Perlack RD, Stokes BJ (2011) U.S. billion-ton update: biomass supply for a bioenergy and bioproducts industry. ORNL/TM-2011/224. Oak Ridge National Laboratory, Oak Ridge, TN, U.S. Department of Energy

Perrin R, Vogel K, Schmer M, Mitchell R (2008) Farm-scale production cost of switchgrass for biomass. Bioenergy Res 1:91–97

Qin Z, Zhuang Q, Zhu X (2015) Carbon and nitrogen dynamics in bioenergy ecosystems: 2. Potential greenhouse gas emissions and global warming intensity in the conterminous United States. Glob Change Biol Bioenergy 7:25–39

Rabotyagov SS, Campbell TD, White M, Arnold JG, Atwood J, Norfleet ML, Kling CL, Gassman PW, Valcu A, Richardson J, Turner RE, Rabalais NN (2014) Cost-effective targeting of conservation investments to reduce the northern Gulf of Mexico hypoxic zone. Proc Natl Acad Sci 111(52):18530–18535

Rosas F, Babcock BA, Hayes DJ (2015) Nitrous oxide emission reductions from cutting excessive nitrogen fertilizer applications. Clim Change 132:353–367

Searchinger T, Heimlich R, Houghton RA, Dong F, Elobeid A, Fabiosa J, Tokgoz S, Hayes D, Yu T-H (2008) Use of U.S. croplands for biofuels increases greenhouse gases through emissions from land-use change. Science 319(5867):1238–1240

USDA Economic Research Service (2011) Fertilizer use and price. http://www.ers.usda.gov/data-products/fertilizer-use-and-price.aspx

Wang M, Han J, Dunn JB, Cai H, Elgowainy A (2012) Well-to-wheels energy use and greenhouse gas emissions of ethanol from corn, sugarcane and cellulosic biomass for US use. Environ Res Lett 7:045905

Index

Printed in the United States
By Bookmasters